AGAINST THE WORLD

BY CEZAR MARCILIO

Revision

Jessica Halana Cristo Cardoso

Felipe de Araújo Amorim

Book cover and illustrations

Cezar Marcilio

ROBB ENTERTAINMENT
ROBBENT.COM

ROBB PUBLISHING
BOOK.ROBBENT.COM

FACT:

Some of the places and technologies in this novel are real, but some of their features have been modified in order to adapt to the narrative, which takes place in a dystopian future.

The last enemy to be destroyed is death.

I Corinthians 15:26

PROLOGUE

Sunrise slowly let nightfall come up, but an orange sky still covered the Sims' family property as a frenetic soccer competition happened. Playing in all positions of the game, young Antony, six years old and pure energy, ran among his cousins and brothers with a single goal — to just kick the ball, no matter what side the ball would go. All his relatives on the soccer field were around ten and twelve years old, and everyone was being watched by a small crowd of middle-aged men and women of the Sim's family. The field seemed small in the middle of the immense green area located behind the luxury Sims' mansion. This recreation area and its surroundings were wooded and extended all the way to a lake nearby.

Everyone belonged to the same family, everyone was enjoying the moment, playing, drinking, smiling, telling jokes, and supporting their kids. Some drank lemonade as others drank beer and happiness seemed to prevail in this late afternoon. It was one of those delightful moments that nobody would wish to end — especially little Antony, who had been admonished every moment by his worried mother, Angela.

"Don't run so much, Tony. Be careful with your heart," the caring mother warned unsuccessfully. She knew very well about her unique son's health problems, and she constantly lived in a complicated dilemma between letting her Tony act as a normal kid or protecting him from something bad that should never happen.

"Oh God, I'm so worried about Tony. I never see him running this way, if he was calmer, I would be calmer too. Oh God, imagine what can happen if I'm distracted for a few seconds," Angela chattered to her older brother Peter, who was always relaxed and always holding a bottle of beer.

"Relax, Angela! You know, if God existed, he would already have taken Tony a long time ago hahaha," Peter claimed sarcastically with a soft beer breath.

"Hey...watch your mouth! Don't joke about Tony's health," Angela poked her brother as she unconsciously held the crucifix around her neck, "Only God knows about my worries and pain."

"Calm down, sweetheart. I'm just joking and —"

"We must not joke about God's work," Angela looked angry.

"I can't argue with someone I don't see. Anyway... it's getting dark. It's better to end the kids' game for today," Peter moved away from his sister as he called the kids to stop the game. Despite her feelings, Angela knew very well what her family thought about her beliefs.

Other relatives followed Peter and called their kids back to the mansion — ending one more lovely day with the family. Of course, it wasn't an easy job making kids full of energy stop playing and, as always, the first call was useless.

Antony and the other kids played like tomorrow didn't exist — like the call of their relatives was inaudible.

He was the youngest among his cousins and seemed small compared with children of the same age. Antony had long curly black hair and a pale, thin body that highlighted his fragile appearance, but not his energetic attitude.

During the last game, one of Antony's cousins, Clifford. crashed brutally with his sister Christine. Clifford, who was famous for being brave and always taking games too seriously — never playing just for fun —, kicked the ball so hard that it was launched into the wooded area behind the field — where the woods were dense, making this area gloomy before getting to the lake. Antony ran toward the trees, shouting, "let me get it, let me get it." His reaction was so quick and typical for a kid that nobody paid attention to him, except his mother.

"Go fast, *abortion leftover,"* Clifford screamed to his cousin and almost instantly was hit with a hard slap from his mother.

"That hurts, Mom," Clifford made a crying face.

"I already told you that's a very impolite nickname," his mother scolded him.

Suddenly, little Antony vanished among the trees.

"Oh God!" Angela whispered, looking at the dark side of the family's property.

Antony ran against the shadows, dodging the bushes and branches without losing the smile on his excited face. He didn't notice that the noise of the crowd became whispers far behind him till he got to the lake. Around there the small forest is open and without trees. The ball was easily located close to a big rock — slightly smaller than Antony — as the sky became darker. He quickly went toward the ball and crouched, wrapping his little hands around it. However, he was petrified when he felt a cold wind hit his skin. A cold wind not from the weather but from someone who landed on the rock in front of him.

He was shaking, sweating, and overpowered by sudden fear as he held the ball tightly. Slowly his eyes moved up as his mind struggled to process what they were seeing.

"Who... Who are you?" Antony stammered and his voice almost failing to express the words.

"You don't know me now, but you'll know me soon," she said as she was crouched on the rock moving her wings slowly — wings made of gray feathers with black stains.

Her skin had a dark beige tone, almost yellow, and her face expressed the seriousness of a prosecuting attorney, ready to claim their evidence, despite looking like a thirteen-year-old girl wearing a short, tattered dark dress. Her long, oily, black hair covered her shoulders. Her body had black and blue spots and stripes — like body paint of South American Indigenous people. But what impressed Antony the most were her eyes. They had an incredible light blue tone, like two pieces of a clear sky, with narrow pupils like cats' eyes. Her gaze seemed to hypnotize little Tony.

"How? I'll know you?" Antony was terrified and had a strong urge to run away and shout out the name of his mother. But something inside his subconscious seemed to want answers, so he stood, holding the ball tightly in his hands.

"You will understand it in due time. You will know that everything depends only on you," she said in a calm voice.

Antony suddenly dropped the ball and started to complain.

"My head... my head is itching... itching a lot and hurting... AAAAHHHH."

His head started swelling up as his hair fell, and he couldn't close his bulging eyes, which were dilating increasingly. His head, covered in dilated red and black veins, became the same size as his body as he raised his hands to sky — begging God for help or simply trying to balance his big head.

She just watched this freak scene without expression, without movement, as if it were the most normal thing she had seen in her life. She stayed crouched on the rock till the top of Antony's head exploded like a balloon, and from there, a huge mass of smoke and white lights emerged. Antony's desperate face cried a horrific guttural song while the smoke and lights formed a whirlwind from his head that reached a few clouds up in the sky — far away from there.

She moved her blue eyes up toward the sky, keeping her cold expression.

"You will understand everything in due time," she said.

Pain, fear, horror, and doubt.

Every dammed feeling together in a whirlwind raped his mind, then Antony closed his eyes.

Silence, only silence, and when his eyes opened, there was nothing more. Nothing but a strange smell — a mix of cigarettes and expensive beers.

CHAPTER 01

The year was 2084. A corporation led by the Sims Family, specialized in social medias, inaugurated not just the greatest enterprise of the world but a brave new concept which had quite literally made the masses change their ways of facing and understanding *life after death* — something that would separate humanity between before and after its creation. It was called Cloud.

Since the beginning of the 21st century, people around the world had made it a habit to share their wishes, what they liked, what they didn't, their feelings, what they loved, where they were, arguments, ideologies, good moments, bad moments, they had put themselves in a virtual net, but they hadn't realized it, except for one man, Allan Sims. The patriarch of the Sims Family — the man who pushed all his relatives to work in new concepts of technology associated with current population — had realized people downloaded their lives into systems that host user files and social medias. However, it was not this habit that drew Allan's attention. What truly mattered was the essence of this habit.

For years people had gradually been exchanging real life for a virtual life, where they had a pseudo-control — a place where they could talk to and build relationships with whomever they wanted, a place where they could say whatever they wanted to say. Also, on the internet, there were algorithms — a complex sequences of codes which detected humans' wishes and motivations every single minute. Then these codes showed people just what they wanted to see — imprisoning the population in their own pleasures. Reality became just a complement to this virtual world or a place to talk about the virtual life. All the while, being so easy that any child could understand.

Also, Allan had realized, in the course of time, a large percentage of the population became atheists. More and more people changed their religion concepts for self-help books or coaching lectures, making churches become empty — not just churches, but Hindu temples and all sacred places that had been created with the purpose of caring for people's souls. Furthermore, another large percentage of the masses exchanged these places, where they could reflect about their lives, for the momentary dopamine release triggered by short funny videos, emojis, pornography, games, poor arguments about ideologies, and tons of addicting stimuli available on their hands through their phones. The global network became a miscellany of information where people got immediate relief and a way to unload their feelings. Although, at the same time, suicide and depression cases grew increasingly, and nobody liked to talk about this silent evil that was expanding... or this matter was purposefully kept secret.

Through this unending supply of information in which the population was adrift, governments and powerful corporations realized that they could induce the opinion and the will of masses according to their wishes. People who were stressed, distracted, confused, unhappy, busy in a world with excess information — most of them useless — could be easily controlled. Through the internet, these powerful people could keep the population stressed, angry, depressive, afraid, and people in these emotional states could be manipulated by determined information. Who was controlling this information through which the population felt like they were randomly flowing around the internet? Just a small group of powerful people — who enjoyed their high positions in big corporations and governments — who had an immense desire to control the world. They were like invisible priests leading their followers who thought they had free will to decide what they wanted — but who in reality didn't have any. In this strange world, invisible leaders guided an immense crowd of blind devotees — who only felt able to see for themselves. However, one thing couldn't be controlled yet. Death.

Death was the final frontier between these few powerful leaders and the population. Death was the only common thing among humanity, the only thing easily understood by everybody, the only thing that made people doubtful and afraid — still Allan Sims had an audacious idea... and this made controlling the fear of death possible.

What is Cloud? Allan always started his business speeches with this question. *Cloud was a new heaven, the other side of life, an amazing place where everybody will live one day, a world full of everything they liked.* Allan Sims developed a system able to transfer all consciousnesses and memories of people instantly after death, to a virtual world that was built based on the algorithms spread through the internet. These algorithms — which for years had checked the wishes and desires of the population — were used by hundreds of programs to build a beautiful virtual society in Cloud, where crimes, poverty and pain didn't exist.

Every single day we are building our own heaven: Allan contended to his shareholders. *The only thing left to offer to people is the certainty of knowing where they go after passing away... and keep in mind that death is the only certainty everybody has... anyway, I can see that as a business opportunity. If society can choose what they consume, where they can go, their friends, their meals, games, movies, ideologies, beliefs... they can choose their heavens and how these should be... and the algorithms can help us in this enterprise. Moreover, we can offer virtual trips for people to see their future homes and clarify to them that they can visit their loved ones, who already passed away, at any time with a simple computer program or app. We're going to build a freeway between life and the world beyond death.*

Damn madman, this is the way Allan was called as he gave all the justifications and reasons beyond this supposedly very crazy idea. But day by day through his incredible power of persuasion he convinced shareholders, businessmen and politicians to privatize the human faith — until he succeeded in convincing all his business partners that *life after death* was a great market niche.

The *Iceberg Man* was how this great and gifted man was nicknamed because of his pale skin and big stature — a fat and strong body at the same time. Bald, with deep black eyes and a strong voice, able to lead a crowd to a civil war or call many animals in an African savanna — Allan was already one of the richest men in the world when he had the idea of Cloud.

After he got the trust of many powerful people, a frenetic rush to build a complex and high-tech system to transfer the consciousness and memories of people to Cloud was created. Chips installed in people's brains when they were born or in other periods of their lives would download their minds to Cloud through powerful transmission towers and satellites around the planet. When someone died, these chips would automatically transfer the human consciousness and memories to Cloud, where people could enjoy the perfect life they always wanted. In the beginning of this system implementation in 2084, public and private healthcare systems joined to implant these chips in people's brains. In the following year, everybody, minutes after being born, received their chips in the labor room. Some years later everybody already had chips installed in their brains. Despite the impossibility of chip extraction, an insignificant number of people didn't agree to it.

The idea to install chips on everybody around the world and replace heaven with a virtual world created by humans was shown in different medium some years before its implementation, like sci-fi movies, games, soap operas, podcasts, books, among other things. It was very important that the population got used to this novelty until Cloud became a reality and started being treated as the unique solution for humanity's fears. *Imagine, after you die, you have the certainty to wake up in an amazing perfect society. Also, you are free to be visited by your relatives and friends. That's awesome.* So, people weren't afraid of death anymore, and religion was abandoned, because the population didn't have any reasons to try to understand life after death — a big corporation had already solved this dilemma.

Over the years more and more investments financed Cloud and its perfect virtual world as reality was forgotten by authorities and leaders. Impunity, crimes, unemployment, arguments, and protests about who had the *best ideologies* — scuffles between opinions of left and right were the easiest way to distract people from other important matters.

The world, full of sad people — addicted to emotions supplied by the internet — became a place where everyone wanted to go to Cloud, and many of them even preferred to kill themselves to accelerate this process. *People live in a true hell hoping to go to a virtual heaven one day,* a famous journalist said one time.

Once Allan remembered this time when he was watching the news and a wise professor said: *Allan's creation transformed real life in a Dante's purgatory while a fake world became heaven*. If he felt satisfaction or not by his invention, nobody knew, but certainly he became one of the most important men and an honorary member of a global organization called WPU (Worldwide Peoples' Union), a more affectionate name to the former World National League — basically a team of government leaders and their bureaucrats who wanted to decide the course of humanity. This organization kept close relations with the Sims Family, and most of its members were business partners of Sims' Corp.

Sims Corp has crossed the decades as a kind of monarchy where just member of the family has administered their business including Allan's grandchild, Antony.

Transferring people's minds to a huge virtual place became easy because it was easier, through some programming codes, to build millions of perfect and comfortable virtual places for the population than for construction companies to spend a lot of money building homes and all kinds of resources to avoid polluting the planet.

The year was 2184, twenty-six years after Allan Sims passed away and the birth of his grandson and heir of Sims Corp, Antony. Cloud's population had reached four billion and it was growing more and more every single day. This achievement was celebrated by everybody who was involved with Sims Corp, except for its newest member, Antony, who had decided to abandon his family and distance himself far away from its business two years ago.

All TV shows, news, and reporting sites had publicized the gifted boy: *The pride and future of the Sims Family, who left the most important position in the richest corporation in the world.* The media also remembered that the emblem of the Sims Family is the symbol of infinity made out of algorithms ones, and zeros.

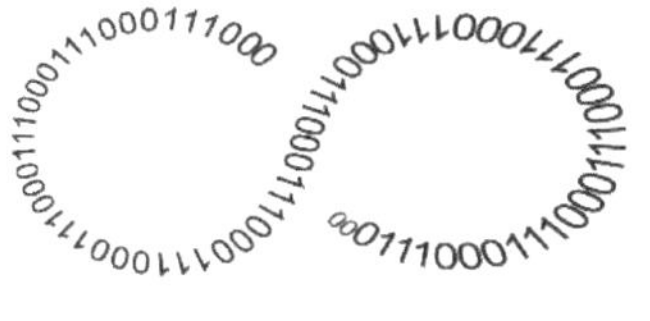

It's apparently not so infinite was a joke made among some journalists. Actually, nobody knew exactly where Antony was living. Some months ago, he launched a book called *Cemetery of Souls* that made a lot of noise because everybody thought it could be an autobiography, and that Antony would explain the reasons that made him leave his family and powerful position there. However, it was just a book containing several weird tales and quotes about angels and how life after death could be. It was a big disappointment to many journalists who were expecting great revelations but just got a difficult-to-understand book that reached an insignificant sale.

Despite other family members keeping the leadership roles of Sims Corporation, two mysteries still intrigued everyone. The first, and least talked about, was: *Why did Allan Sims, after passing away, prefer a true death rather than living in Cloud, his own invention?* The other, and most discussed one, was: *Why did Antony keep himself isolated? A guy who literally had the world in his hands*. News and gossip TV shows loved to create theories about that — like the last time Antony was seen, about midnight, walking down a street of the center of Brussels, Belgium. Nothing seemed strange, except for his appearance, as he looked very drunk.

CHAPTER 02

Despite ninety-eight percent of Earth's population not believing in any religion, the Gregorian calendar, introduced by the Catholic Church, was the one still in use. So, the year was 2184, chosen for convenience rather than religious tradition, and the city where incredible changes in the world were about to happen was Brussels, Belgium. Gray clouds extended across the sky, and a light drizzle fell over the center of the city, making this environment, in its entirety formed by ancient buildings, look more and more depressive, as dawn rose shyly. The tallest building was visibly the most modern, with a wavy architecture, big mirrored windows, and a heliport on its roof. The last three floors had a different design, which meant this whole part of the building belonged to a single person. Despite its futuristic design, this building annoyingly contrasted with the rest of the urban environment, where buildings varied between gothic and neoclassic architecture styles.

In the living room of the penthouse apartment, a twenty-six-year-old guy slept awkwardly on a large sofa. It was around 10:00 a.m., and in front of him, several empty beer bottles and a tablet stood on a glass coffee table. The morning seemed night as Antony, clearly hungover, was asleep.

His pale, thin body, dressed in a white tank top and baggy jeans, looked fragile, like a patient in a coma. However, what really brought out his weak appearance was a big surgical scar in the middle of his chest, which meant this guy had suffered from health problems his whole life. Otherwise, he was tall and had wide shoulders, like an Olympic swimmer but underweight for an athlete. Despite his age, his slim face looked like it belonged to a main actor in a teen movie — handsome but with a somewhat depressive countenance.

Suddenly he widened his deep black eyes, like he had just escaped the worst nightmare of his life. Antony looked very scared and breathless as he quickly got up. However, the fright was momentary, just like it had been every single morning. Then he kept seated, feeling the fright fading into a headache — effects from the hangover.

"Another bad dream, my bright boy?"

The question came from the kitchen, next to the living room, in a sweet and calm voice. It was Nivedita, the housekeeper of Antony's apartment, who has worked for the Sims Family for years. Since Antony's mother passed away, she had become a kind of stepmother to him, always careful and willing to care for her "adopted" son. She was decedent from a traditional Indian family and had come to Europe in her youth. This lovely short lady was in her sixties and had a simple routine between her life with a little Indian community in Brussels and her work at Antony's apartment. She was a typical Indian woman who had big black eyes, brown skin, and long straight hair with some gray locks. Her clothes never varied from a simple long dress and a colorful pashmina on her shoulders.

For a long time, she was the best friend of Antony's mother, Angela, despite their cultural differences, mainly about religion. Nivedita followed the Buddhist philosophy since she was a child, and Angela was a zealous catholic. Angela probably asked Nivedita to always stay close to her son — through good and bad moments — just in case something bad happened to her. Whether this request had really been made or not, nobody could confirm, but Angela's presentiment had become reality. She had been a victim of a sudden cardiac arrest two years ago; since then, Antony had never been the same person.

"You came earlier today! It's Saturday. Didn't you have parents to visit or something... WOW... my head." Antony could barely hold the weight of his own head."

Nivedita smiled as she approached, cleaning some little statues on the glass shelf behind Antony.

"It's Sunday, my bright boy, and it's 10:15 a.m. Also, did you forget my only family here in this country is you? Well, I still have some friends in this city I could introduce to you. Because I am sure you drank all these beers by yourself last night."

Antony got up slowly as he combed his long, oily, curly black hair. Then, by instinct, he walked up to Nivedita and opened his arms to hug his stepmother. An affectioned kiss on her forehead made his momentary bad mood vanish.

"I'm sorry, Nive, very sorry... I always forget the person who needs to have people over in his house is me. But I don't see the problem in having beer bottles as best friends."

Antony's mind had always worked in a differently since his childhood. Complex calculations, math formulas, sequences of algorithms, and computer systems were easily understood by his super developed neural network. His fast mind was also able to quickly analyze all expressions and attitudes of any person, making it possible to predict their following actions. This way of facing people and situations made Antony a confident guy with an oscillating personality — expressing his ideas and opinions firmly and always with a pinch of sarcasm. However, a developed brain didn't mean good emotional control. When it came to his relationship with his family, rage and harmony seemed to be always in conflict — a mind that could calculate the hardest quantum physics problems in the world but a heart would sometimes cry in solitude.

Antony slowly walked away from Nivedita and leaned on the sofa as she kept smiling and cleaning the statues on the glass shelf — all of them were miniatures of saints and gods from different mythologies and religions.

"No worries, my bright boy. Sometimes I prefer staying here with you than with my friends and relatives. It's more challenging," Nivedita joked. "How about your last dream? Was it the same thing? You're playing soccer when you were a child... you see an angel then your head explodes. Is that it?"

"Yes, the same damn dream that you've already memorized. If my mind could be more creative, I would drink less. Have you made coffee?"

"Of course, I have. It's better you drink a little before you see your mother. No mother would like to see her own son hungover, despite most of them pretending they don't mind."

"My mother?" Antony looked startled. "Oh, yes. Today is Sunday. The day to go to church. Oh, I almost forgot," he checked the time on a wall clock in front of him. "Humm, but it's not 10:30 a.m. yet. I have time enough... ooooh, my head. If this damn dream didn't disturb me every night —"

"I have always said this to you. This same constant dream is a message...a message that you just can't decipher," Nivedita spoke calmly but sometimes sounded like a severe mother. "While you can't decipher it, it will continue. And staying isolated from others? Humm... I think it can't help you nowadays."

"Well, I wrote a book like you suggested. I described everything I think about, life, death... you know... the craziness that this society has become. And nobody liked it. That was a true sales failure. Nobody cares about what I think."

"It's not a question of making money or people understanding you. It's about embarking on a journey within yourself. You and your family have money enough to buy this city."

"Actually, my family bought this city some years ago. It was the only way to build this ugly building here." Antony looked more awake and more sarcastic too. "Also, I admit that writing a book was a good experience, better than diving in Malaysia, piloting a helicopter over an erupting volcano in New Zealand, attending a motorcycle championship in Italy and —"

"Hunting crocodiles in the Nile River with a rifle, riding a jet ski, going to space to fix a satellite... I know, I know all your adventures," Nivedita completed, "you've already lived a lot of adventures despite your young age. But did any of it work? Are you feeling better now?"

"No, it didn't," Antony was a little disappointed and confused. "And I'm feeling the same shit as always. You know... these nightmares, my family, all the shadiness of Sims Corp. Also, the brothels and pubs around here that never satisfy me."

Nivedita continued in a loving voice. "What I mean, as you have said, is that the world is crazy and there is a message from a superior power... I guess. This power is using you to spread this truth. It's a matter of time until people understand you, my bright boy."

"You and your ideas... you talk like my mom."

Nivedita held a statue of Siddhartha Gautama, better known as Buddha, and carefully cleaned it as she talked to Antony.

"Look at Siddhartha. He was a prince who belonged to a rich family... like you and he left a life full of luxury to live alone because he had a strong necessity to understand humanity. Pain, old age, sadness. But he almost died because his own will wasn't enough to satiate his need for answers. He kept himself so isolated that he forgot to take care of his own body until, suddenly, a person came to him and offered him a meal. Thus, he understood that he couldn't reach his goals alone, not in that moment. You see something wrong in our world, but you don't know how to solve it yet. For me you're like the prince Siddhartha, my bright boy."

"Why do you insist on telling me this fairy tale every week?"

"Why does your mind insist on the same dream almost every night since you were a child? That's the question you should ask yourself. Also, don't call it a fairy tale. I always tell you this when you talk this way."

Antony had a shy smile on his face as he slowly went to the sofa and sat there. He held his tablet and started to check something.

"Well... sorry again. To be honest I love when you talk like a coach, especially about this prince and that other fairy tale that you always tell me about... ah... the man who was crucified between two robbers..."

"It's not a fairy tale, Antony... and remember how your mother loved this story," Nivedita looked annoyed.

"Hey, take it easy, I'm just joking," Antony's voice ranged from sarcasm to respect and laziness. "I don't understand why you and mom insist on these tales that were used to educate kids in ancient times. Anyway, back to your first *coaching moment*, you know... I think this superior power, as you said, must be looking for a prophet who drinks less than me. Besides, I don't want to leave my comfortable life like Siddhartha. I just want to get rid of these nightmares and —"

"You don't like what your family has been doing to the world all these years," Nivedita completed the sentence.

"Yes," Antony breathed out, "maybe... if I could do something or... hey... look at this!"

Antony seemed surprised as he watched something on his tablet.
"What's the matter?" Nivedita looked curious.

"I got a message from the website where my book is available for sale and... Wow, now I'm on my account at handfriend.com and... WOW... Jesus."

"Hey, when you get scared you remember his name," Nivedita smiled sarcastically as Antony cast an indignant look at her.

Handfriend.com was the website of a company that administered several businesses around the world. Actually, it was responsible for producing everything related to culture. It began as a simple website that sold books but quickly started buying other popular sectors — movies, cartoons, documentaries, songs, shows, online schools, and courses. The list of available services seemed endless, and these services had specific features capable of influencing the population's beliefs and ideas, as chosen by WPU, its main investor. Thus, Handfriend Corp controlled what the population could think and what it could not.

One year ago, Antony published his book, *Cemetery of Souls*, had been an unsuccessful kind of therapy to him — until that moment.

"Oh... I hate this website, but... anyway, it's the only way to sell books, and my family never thought to buy it."

Antony didn't blink as he checked if the information on his profile was right.

"Hey, don't leave me waiting. I'm getting apprehensive," Nivedita hit Antony's shoulder with a cleaning cloth.

"My book. In the last months, I sold nothing, but now... To be more exact, in the last 24 hours I have sold... 8,572 copies. It's insane."

Nivedita looked intrigued too.

"Maybe something happened with your family. Don't you know anything? Have you seen the news? For people to suddenly buy anything so fast, something must have reverberated recently."

"I have no idea. Also, I've distanced myself from the news and my family a long time ago," Antony appeared confused, but he didn't express strong desire to find answers to his doubts.

"Anyway... well, I have to meet my mom at 10:30. How long do I still have, Nive?"

"Three minutes."

"Cool... I plenty of time. Enough to have a cup of coffee before I go."

CHAPTER 03

It's 10:30 a.m. A red chaise lounge, like the ones used in psychiatric offices, was set close to the biggest window, on which droplets formed as it drizzled outside. The sky was still gray and sad as water from the drizzle slowly rolled down the windowpane — blurring the vision of an already somber landscape. Antony lay down on chaise lounge and put on his forehead a device that looked like a thin crown, named Neural Connection Between Cloud and Reality Mechanism, but this name was so hard to memorize that people decided to nickname it just Headgate. This device was sold very cheap — almost free — on any market or online store. Thanks to this device's system, created by Allan Sims, anyone could lie down and sleep as their mind visited Cloud for a few hours, until its body felt like waking up.

Antony took his phone and typed so quickly that his fingers looked like stains on the buttons — a skill he acquired while working in the Engineering and Technology Department of his family's corporation when he was younger.

Antony's family started noticing his abilities when he was four years old. A great capacity to solve complex math calculations and the ease of understanding all kinds of articles and books got him into university when he was nine years old. Of course, his family, and especially his grandfather, hired the best professors in the world to push Antony's cognitive capacity increasingly, but they got to a point where even they couldn't help anymore. His high IQ and his super-developed synapses were incomprehensible to any scientist. Celebrity gossip websites and TV shows speculated that the Sims Corp heir had been created in a laboratory. His mother insisted he was a blessing from God, and his grandfather said nothing except that *he is the right person in the right place.*

Maybe it was just a coincidence that a super gifted boy was born to the family that owned the most powerful corporation in the world.

Antony was the promise who would make the Sims Corp an institution bigger than any government. In the beginning of his adolescence, he enjoyed life with a kind of *God status* — travels, parties, meetings with the most famous people in the showbiz. He, better than anyone else, knew how to reconcile his work with the pleasures the world could offer — a guy who was adored by teenagers because he showed them there was a life that could be enjoyed and that after death there would be more, much more pleasure. Antony was a symbol to show that life after death could be controlled because he controlled it.

An intelligent and extroverted guy controlled what many religions had tried to control... until his mother passed away. Then the *golden boy* of society mysteriously changed his mind, and Sims Corp started being led by his egocentric relatives — who never had his brilliance.

"It's 10:31, I have to go," Antony whispered as he left his phone on his belly and relaxed his arms.

"Ok, here we go!"

After a few seconds, Headgate made a low click sound, and Antony closed his eyes. Then he seemed to be in a deep sleep.

CHAPTER 04

Kids ran and played like there was no tomorrow on streets decorated with colorful flowers and flags. It was *Spring Day*, a day of confraternization, exchanging gifts, and playing games. A day in which it was impossible to find someone who wasn't smiling — having fun seemed obligatory and happiness was contagious. The sun shone intensely up in a clear blue sky, the air was fresh and there weren't cracks or flaws in any of the houses. Green trees and flower bushes were cared for by happy adults in front of their beautiful homes — homes which seemed designed for a fairy tale book. All of these homes were open, and people of all ages came and went carrying meals and gifts. Many women walked joyfully sharing gossip and talking about TV shows as teenagers had fun playing baseball, football, helping their parents in the gardens or just flirting.

In this place *smiling* seemed to be a kind of *obligatory adornment*, except for a guy walking down the street — Antony. He arose from a light that flashed so fast that nobody noticed his arrival. He kept on his way with a dissatisfied look, wearing the same used clothes he was wearing in reality. A well-worn blue nylon jacket, jeans, and white sneakers.

Welcome to Cloud again, Antony. Better than perfect, like my grandad used to say. This place is sickeningly happy, he thought.

Everything described about this place up to now was inside a huge software created for the most modern computers in the world, and all people here had already died. Just their memories and consciences lived here, no pain, no famine, and no violence. Their bodies were probably cremated or frozen — depending on the plan they and their families got when they were alive.

Antony liked to call this place *The Biggest USB Drive in the World,* as people in real life named it the new heaven or just Cloud. Actually, Cloud was an immense building located in the middle of the Nebraska desert, built in a modern design based on the Pentagon in Washington, but bigger and with more security because this place housed thousands of computers where billions of peoples' memories and consciousnesses were stored. A complex AI system and thousands of programmers, IT technicians, cyber engineers, and scientists monitored everything. Also, there were isolated parts of Cloud named *Mindcages*, where the minds of those accused of crimes were stored, far away from the good people who lived in Cloud — like a virtual life imprisonment. Nevertheless, they were watched by law professionals who evaluated their behavior, and if they were found innocent, they would be transferred to more comfortable places, like where Antony was walking — where there weren't bad people, and only happiness and good vibes reigned.

"I can't believe I helped creating this damn place full of empty people," Antony murmured.

He kept talking to himself as he went on his way until he got to a beautiful white wooden house surrounded by bushes with pink flowers. On its porch, a pretty lady in her sixties was sitting down on a bench, reading a bible. Her long, smooth silver hair in a ponytail and her rosy cheeks accentuated her black eyes on her angelic face. By instinct, she looked up from her reading and gazed ahead of her.

"My son," she whispered and smiled.

Angela got up immediately and walked toward her son, holding her bible tightly in one of her hands. She wore a long dress with a floral print and knitted coat that made her look more like Antony's grandmother than his mother.

They hugged in front of the house, and Antony smiled for the first time since he arrived. He felt very comfortable and relieved, like a baby on its mother's lap, despite looking bigger next to his small and thin mother. In these few seconds, surrounded by his mother's arms, he imagined how a normal life in reality with all his family together in a big weekend lunch would be. He knew it was impossible, but his mother's love had always made the spark of this far away dream burn.

"I missed you, my son."

"I come here every Sunday. There is no reason to miss me so much."

Slowly, they released each other.

"One minute late? You can't arrive at any time you want just because you built this place," Angela admonished him without losing her lovely expression.

Antony looked relaxed as he smiled, then they walked arm in arm down the street. From this moment on, Antony seemed to appreciate the place and the excited folk around them — just a little more.

"You look so thin. Have you been eating well? Or are you still on your *beer diet*? I pray every single day that you'll change your habits. I can smell the alcohol on your clothes. You're the main representative of our family corporation. Do you know that?"

Despite all the technology that had already been created over the centuries, the behavior of moms hadn't changed at all. Antony always talked to his mother with humor, patience, and his typical sarcasm.

Angela looked like a fragile lady, always talking calmly and without interruptions. But hidden behind her angelic face was a woman who had strong opinions, which hit hard anyone who dared to get in an argument with her.

"Mom, I'm ok. No worries. Beer has a lot of protein, I guess... Anyway, this smell you're talking about is actually a computer program that I'm gonna turn off, so you don't complain next time. If I don't drink next time, of course, I'll remember it."

"I told you to stop explaining everything here as if this place was just a computer program. You have continued the amazing work of your grandfather. Look at these guys. They are happy, completely happy —"

"They are codes inside a memory, monitored by an AI and men who want to keep their jobs and ensure their retirement. You know, these damn men watch everything that goes on here, even when you're in the bathroom... OUCH —"

Angela slapped her son's chin, but her face was still lovely and her voice calm.

"What about that? Was that programed too? Naughty boy! This place was made by your grandfather to help people, and you were born with this same brilliant capacity. If the Lord gave you a gift, He knows that you'll use it to help others."

Antony looked a little downcast at his mother's words; nevertheless, they kept talking and walking along the street full of happy people. Several kids passed by them on bikes, skates, or scooters, some of them with flowers on their heads. Everything in this walk was so colorful and full of life, but Antony didn't pay attention to what was around him. His coldness regarding Cloud made him see just codes and algorithms. At the same time, he felt so sad seeing the number of children. Even though he knew that some of them were there just to visit their parents who had passed away, he knew very well that there were kids who unfortunately had died early and had been adopted by families who already lived in Cloud. But there were also many adults who decided, after death, to bring their kids over through euthanasia. This last situation disgusted Antony, especially since this was a choice granted to the families and totally legalized by WPU. Moreover, a particular detail Antony easily noticed about these kids was that none of them were under ten years old.

"You know these kids are living a fake life," Antony continued, "they don't bleed, they don't have reasons to cry or frustrations. Also, their first sexual experiences will be watched by the men from the IT department —"

"At least, thanks to your grandad, they'll have the opportunity to grow up and have this kind of experience... and others as well. All this without having to worry about becoming old like me. This is a blessing."

"I know, I know... every kid here will age until their fifties. Have you heard that the damn WPU wants to decrease this age to forty? In a few years Cloud will be full of stupid and immortal people who laugh at anything and —"

"Hey, son, calm down! You always come here with these speeches about Cloud. Give your mother some attention, please. Anyway, most of these kids are here because they and their parents decided it. They could have come here just to visit their parents… like you… but some of them decided to stay because they were fascinated by this place. This place that you helped build."

"Yes, I know," Antony looked up trying to think of an answer, but nothing came up.

"And what about the others? Uncle Peter, Christine, Clifford, Mathias. You loved playing soccer with your cousin Mathias."

"I haven't seen them for quite a while" — Antony was unable to lie to his mother, no matter the circumstances. "Nivedita is my family, that's it. The others come looking for me only when they need to solve some stupid problem in Cloud which I can always do on my phone."

"Ah… my kids, the more I pray that you'll become adults, the more you act like kids. An old woman out of here and another old woman in here can't be your only companions. You should meet people more like you, who are the same age."

"The same age is easy, but like me… I think that's impossible. The problem is the world won't put up with two of me… Hey, here we go."

Angela and Antony approached a church located in the middle of a small park surrounded by some high oaks. This building wasn't as big as the ancient Italian cathedrals but not as small as like the ones in the countryside. Its eclectic style was a mix of neoclassic with some baroque features, built with red bricks and adorned with white pediments, columns, and cornices. Massive white doors and windows combined with a white dome above gave a conceptual appearance to this church, mainly by an interesting detail: a sculpture of the archangel Michael with its wings wide open, in the act of drawing a sword from its sheath, stood on the massive porch at the entrance. Below this angel, on the pediment that formed the porch, were written the initials I.G.W.T in roman style.

Every programmer of Cloud knew that this church was a gift from Antony to Angela, who couldn't resist asking her son to adorn it with some special details like the sculpture of Michael and the initials.

In this part of the neighborhood, there were fewer people — actually, most of the time it was empty.

After stopping for a moment and making the sign of the cross, both went inside, still arm in arm. Angela was the only Christian of the family; she always talked about her faith, but only Antony listened to her teachings. Despite not following any religion, his mother's teachings floated around in his mind.

The church was empty, as usual since it was built. Its walls were adorned with little sculptures of angels, cherubs, and golden details in a baroque style. Surrounding the interior, many candlesticks with red candles lit the environment. Eight heavy wooden pews placed in two rows separated by a corridor lead to a big altar in the back. Beside it, there was a little white chapel — Angela's favorite place. This chapel was adorned with gold details all around it, and in front of it, there was a gray sculpture of a book on a stone stand. This sculpture was completely different when compared to the rest of the chapel's details. Some stone sculptures of saints filled the interior of this church, but the most remarkable one was located in the little chapel, a perfect replica of a sculpture named *Virgin and Child,* made by Giovanni Pisano in the 14th century. This masterpiece showed Mary holding the baby Jesus.

What drew Angela's attention was how incredibly natural the way Mary and Jesus looked at each other was, like they understood how important they were. In front of Giovanni Pisano's masterpiece, the weird gray stone sculpture of a book on a four-feet-high stand seemed to have no connection to the rest of the church. Also, a high-relief symbol on the cover of the stone book seemed to be a mystery — the symbol of infinity.

"I never understood this stone book. It's so ugly, but I love Mary holding the baby Jesus," Angela looked fascinated, "I know it was your idea."

"Well, maybe... I never understood the initials outside. I.G.W.T? It sounds like a cellphone brand."

"Oh, Tony... it's so obvious. One day you'll realize what it means. But... Look at Mary! Look how she observes Jesus. It's like she wants to teach him something, but he's the Lord. There's nothing to teach him...but he seems to be expecting some teaching from her. It's so simple and amazing at the same time. Better than this weird stone book here. I don't understand why you made it."

"I didn't build this church by myself, especially this freak sculpture or... anyway. Well, most of that... yes, but not everything. Mary and Jesus are cute, I guess. But this stone book, hmmm... I'm not sure. Maybe some boot-licking programmer of grandad's who wanted to honor him. Putting this symbol on the cover of a book in the church could be comparing the corporation to a holy bible or something."

"It doesn't seem like an honor. You could take that away from here, couldn't you?"

"I probably couldn't. Like I just said, Mom, I didn't build everything here. It's a work that was done by thousands of hands. Let's respect the work of other programmers. Also, it's just a detail of the place. A place nobody ever comes to."

Angela looked down and touched the stone book.

"Unfortunately, you're right. Nobody has come here except you and me. I still remember your grandfather talking about the symbol*: infinity in our hands, humanity finally got its freedom.* That horrible and severe voice! I don't want to remember it... ever again," Angela felt a chill.

"Come on, Mom. You... you don't deserve these memories," Antony touched his mother's shoulder as a gesture of consolation.

"Yes, I know."

Mother and son knelt down in front of the *Virgin and Child* to pray. Both knew what the other was thinking and feeling. This was a ritual they had repeated every single Sunday. What Angela didn't know was that Antony had built this church for her. He really hoped more people would attend this place, but it hadn't happened. Nobody in Cloud would change one minute of pleasure for hours of silence inside a church. That's the reason why Antony accompanied his mother. Outside of Cloud, the situation inside churches and temples was the same.

They stayed on their knees, in silence, with their eyes closed for a long time — just in silence.

***

The church's doors were always open, and the nice weather never changed.

Angela and Antony left the church arm in arm, just like they had gone in.

"Well, it's time to have something to eat. How do you feel about hot chocolate and some cookies? Have I told you that I think you're too thin?" Angela looked excited as she talked.

"Mom, you know, nobody gains weight here. I bet the majority here were obese in the real world, but they paid a lot of money to be inserted here with a fit body."

"Hey, at least you could try to be less harsh about it or pretend to like it here. You and this place are the only things that make me happy, but you can't stay here all the time."

Angela looked annoyed for a while, and Antony smiled shyly as he said to himself, *ok, finally, I understand the lesson.*

"Well, Mom. I think some cookies wouldn't be bad," Antony breathed deeply, "one day among these happy people could make me feel better... I guess."

"Yes, let your body relax out of here and enjoy this... this... like you always say, this big USB drive full of fun. Hahaha..."

They came back to Angela's home as people continued the festivities in the streets. Nothing seemed to disturb this relaxed moment after their visit to the church. Nevertheless, Antony heard something from outside of Cloud. He put his hand on his ear as if he were talking to someone through a headset, even though there was nothing physically there. It was Nivedita talking to Antony from his apartment, but he was the only one who could hear her. Antony looked somewhat alarmed by the information he had just heard.

"That's insane. Is she there? Wow... Ok, just her assistant."

"Is that Nivedita? How is she? Ask her to come visit me sometimes, please," Angela looked excited and a little apprehensive too.

"Ok, I'm going now. It sounds like a call to war."

The transmission between Antony and Nivedita was over. Antony then he tried to imagine why someone so important was calling him. He was lost inside his own thoughts for a while.

"Sorry, Mom, I have to go. Someone from the government is calling me to meet her right now."

"Government? Someone important?"

"Really, really important... and dangerous too. But no worries, maybe I can solve it over the phone in a few minutes. I'll come back as soon as I can," Antony answered, smiling.

"You always solve everything. You were born for this."

"I hope so."

Antony raised his face up to the sky, and the sun momentarily blinded him. He was really worried, but he didn't let his mother notice it.

They hugged each other for a minute, a moment they wished could last for the rest of the day.

"Love you, Mom."

"I love you, my son."

They were the only ones who didn't seem completely happy on the street. Then Antony walked some steps away from his mother. *Time to go back*, he said to the system, which understood the sentence as a code to take Antony's mind back to his body and out of Cloud. One flash of light was enough to make him vanish so quickly that nobody on the street noticed it, except his mother.

Angela slowly walked through the streets among the excited crowd, heading toward her home. Suddenly, she looked up at the sky and was mesmerized by a kite far away, its shape hard to define. Actually, it had the shape of a bird, but Angela saw an angel.

CHAPTER 05

Antony had spent three hours in Cloud, but it seemed like a long day. Antony opened his eyes and quickly took off his Headgate, but the moment he stood up, his body made him remember something. Headache, backache, cramps, dry throat — everything he had forgotten in the last few hours.

"Wow... now I miss the damn Cloud."

Antony stayed sitting down for a moment to remember all the pains his mind had forgotten. His mind was exceptional, but his body wasn't, which had made his former workmates nickname him the *seventy-year-old teenager*. Still, he never helped himself on that account. He always refused to do any physical activities or have a healthy diet. But his worst health-related issue was when he needed to undergo cardiac surgery when he was twenty years old. A weak heart that needed constant monitoring at all times was the diagnosis that made him lose his nickname but gain a big scar on his chest.

Nivedita was really worried when she got close to Antony.

"Sorry, bright boy, but he insisted so much... and when he told me who he works for —"

"Ok, Nive, no problem... you did right calling me... oh, my head... Monitor!"

Antony used all his strength to stand up as a blue light at the top of a big screen in the corner of the living room was blinking. *Monitor one*, he murmured in a harsh voice. A voice-commanded system that controls several functions of the apartment turned on the screen where a man who looked like a mix of an executive and an airplane pilot appeared.

"Good morning, Mr. Sims. Your appearance is not so good today, in my opinion."

"My appearance has not been good since I was born. What does your damn boss want with me?"

The blond man on the video was Mr. Aaron. His square face and distinctive accent evinced a typical efficient German official at work and always ready to carry out orders.

"You'll know when you see the situation. Please, meet me on top of your home," Mr. Aaron spoke slowly with a sarcastic smile on his face, like a soldier who was in control of everything.

"Well... you're in luck. I'm very curious about why your boss wants to spend time with me."

"She's not a boss, Mr. Sim, she's —"

"Ok, ok, spare me your damn accent. I'm going... *Monitor one, turn off.*"

Antony expressed clear disinterest and apathy about going out, but his instincts were warning him that something very wrong was happening.

Nivedita looked very apprehensive as she approached Antony.

"Hey, my bright boy. Are you sure you have to go?"

"Yes, I am... she wouldn't order Mr. Aaron to come here by himself for nothing."

"By himself?"

"Mr. Aaron is her best watchdog. He was calling from an airplane cabin which means he came alone. If he came alone, it's a secret mission. Mr. Aaron doesn't pilot an airplane for fun or shopping. She is probably planning one more devilish mess."

"You always talk about her like she is the devil."

"Poor devil. He is not bad enough."

Antony put on the blue nylon jacket that he took from a hanger next to the screen as he talked, then he put his phone in his pocket.

For a second, Antony was hit with an epiphany and calmed down a little. He paid attention to Nivedita's worried face and hugged his *stepmother* as an apology for his ill-bred way of talking.

"It's probably nothing serious. At least I hope so."

"Take care, my bright boy."

***

At no point did Antony express goodwill as he went up to the top of the penthouse.

The drizzle came down harder, almost turning into rain, as Antony spotted a white Cessna Citation XX landed on the heliport, probably the world's fastest business jet of the world. Its retractable, powerful turbines created a perfect combination between rocket speed and the drivability of a helicopter on the same vehicle. However, what troubled Antony a little was the Germanic federation's symbol on the jet, a symbol that meant only important authorities traveled on it. This freakish black eagle that looked like a dragon had always been intimidating. *Wow, the problem is big... really big,* Antony thought.

"Good morning again, Mr. Sims. It's a pleasure to meet you," Mr. Aaron was standing in front of the open door of the jet. He went through the same protocol he was used to with anyone who had the honor of entering this vehicle.

"No fake polite gestures, please. Since she couldn't come herself, let's finish this once and for all."

Antony got in the jet without paying attention to Mr. Aaron, who kept his sarcastic smile the whole time, all the while closely observing each of his passenger's movement.

In a luxurious and heated cabin, Antony sat down in an armchair by the window, not worrying if he was getting it wet since he was soaked through by the drizzle outside. The white and beige cabin looked like an expensive small hotel room with a fridge, cabinets, and a little bar with several brands of alcohol. However, something inside this place was felt off — which made Antony feel troubled again.

The Cessna Citation XX softly took off from the heliport until it was at a high enough altitude to direct its turbines and get more and more speed, leaving the sad Brussels' landscape behind. Quickly, the gray sky changed to a lighter tone of blue, and up there, Antony watched the city of Liege, as ugly as Brussels.

The current century had been really cruel to most cities of Europe, which had had their heyday in the end of the 20th century after terrible wars. At the end of the 21st century, pandemics, unrestrained immigration, and consequently civil wars made the beautiful cities of the world become huge somber towns — a combination of classic buildings and improved modern condominiums with poor designs. Conflicts, violence, riots... humanity was enraged until Allan Sims, in 2084, inaugurated Cloud, the system that calmed the people down in the following years. The system that controlled the world population in the way that politicians and big corporations always wanted — better than capitalism, better than socialism, drugs, religions, or anything ever created to manage and quiet down society.

Through the jet's small window, Antony observed what his grandfather had done: huge, poor, sad cities where people had worked hard to survive and to be able to afford a good place in Cloud. *One century earlier,* Antony tried to image how just one man's idea had changed everything.

"Hello, Mr. Sims. Nice weather now... how have you been these last few days?" Mr. Aaron sat down in an armchair in front of Antony as he poured a shot of vodka. He was very relaxed and tried to talk to his passenger as if they were old friends.

"Shouldn't you be piloting the jet right now?" Antony didn't take his eyes off the landscape outside.

"The autopilot can manage the rest of the trip, Mr. Sims. But I would like to talk about your book. I got a copy yesterday and I couldn't stop reading. Well, I really enjoyed it. *Cemetery of Souls*, personally, I loved the title and your quotes... wow... simply amazing."

Only Antony's doubtful eyes moved towards Mr. Aaron. He knew this henchman was not as kind as he looked, then Antony checked if his seat belt was tight.

"So, you got my book yesterday? Interesting," discreetly, Antony took out his phone and started typing quickly.

"Yes, the stories about angels who came down from heaven because they don't like how people are staying here on Earth after death sounded very interesting. It reminded me of Cloud, your grandfather's invention. Furthermore, the way they came... out of a huge smoke gateway in the sky or a black hole? I don't remember exactly. Do you think Cloud can draw these angels to earth? Oh... sorry, I have been really excited specifically about this story."

"I think you should go back to your cabin before something bad happens."

"That won't be a problem, Mr. Sims. This jet has the most modern autopilot in existence. Can you feel the stability of it? It's like we are sitting on chairs on a beach."

"Are you sure?"

Antony tapped on his phone and caused a turbulence that lasted a few seconds, but it was enough for Mr. Aaron to get scared and drop his shot of Vodka.

"WHAT'S GOING ON?"

Mr. Aaron had completely forgotten that Antony was a computer genius who was able to control anything with circuits with just a few taps on his phone. It started when Antony was three years old and he got his first phone. It was a special phone, a kid's edition actually, but it was enough for Antony to access the internet and hack game apps, hack into some TV channel for kids, and invade a bank system. At six years old, he was able to create algorithm codes with the same ease that a normal kid played with Lego. Any computer, no matter how simple it was, became a weapon in his hands — like this moment when he hacked into the jet system while Mr. Aaron talked on and on.

"No worries, Mr. Aaron. I already stabilized it. But next time, you'd better be ready to run to your cabin."

"Did you do that?"

"Just let me ask some questions now," Antony sat in his armchair as he stared at Mr. Aaron, who looked very scared.

"Why is this business jet armed with a missile system? That's completely insane. You planned to destroy my apartment in case I decided I didn't want to come with you, I guess."

"Sorry, Mr. ... Mr. Sims, but you're making a mistake."

Antony tapped one more on his phone, causing the jet to experience heavier turbulence. A shrill siren from the alarm system sounded, further increasing Mr. Aaron's fear

"I JUST FOLLOW ORDERS. DID YOU KNOWN THAT TOO?" The confident official was desperate in that moment.

One more tap. The jet stabilized again, and the shrill siren stops. Antony put his phone in his pocket and kept a cold expression.

"It's not a mistake. Look at the size of this cabin. It's much smaller than normal. I realized it when I got in here without imagining why. But a missile system? It's more than insane, it's diabolic. I can't believe your boss was planning to kill me."

"Like I told you, Mr. Sims. I just follow orders and the main order was to take you safely to the destination, but you probably already know this," Mr. Arron's breath calmed down slowly.

"Yes, I know. But anyone would suspect a jet armed with missiles piloted by an ex-military man. Now, go back to the pilot cabin, please."

Mr. Aaron was indignant and frightened when he went back slowly to the pilot cabin. Actually, he was a man who wasn't used to lose his self-control in any situation, but the fear of being inside a falling aircraft made him give it up.

Since he left his family business, Antony suspected everyone, except Nivedita and his mother. The lady he was going to meet was one of the people responsible for bringing about this mistrust within him, not because she did anything bad to Antony but because of unethical business she had with his relatives. Moreover, Antony found some wiretaps hidden in the passenger cabin when he was hacking the jet system — which made the situation even more mysterious. He just turned them all off.

She would never dare to order Mr. Aaron to shoot missiles at my apartment, this idea still buzzed in Antony's mind. Maybe she knew Antony would discover it — which made him accept to meet her. Anyway, Antony had to continue with this kind of game to discover what was happening.

The Cessna Citation XX kept its route with impressive speed over the city of Koblenz, another place deteriorated on account of bad contemporary political management, like Liege. Antony watched this other city sadly and wondered how society was able to forget its traditional aesthetics, beauty standards, and change them for abstract street art, paintings and sculptures that needed a lot of psychological and sociological explanations to make sense and justify their ugliness. Even the monument of Emperor William I — located in Deutsches Eck, between the Rhine and Mosel rivers — was changed for a conceptual art installation made out of colorful twisted pieces of steel. Angela had always taught her only son about ancient Art, no matter how hard it was to find it on the internet, as if the world had wanted to forget its own history or a few people wanted the majority of the population to forget its past traditions.

Some minutes later, the Cessna Citation XX approached Frankfurt. Unlike the last city, it was very modern, organized, and clean. It was actually one of the cities in Europe where the richest families of this continent lived. Antony's habit of criticizing everything he saw never stopped. In his mind, the lack of organization in most cities and the excess of it in some others, especially Frankfurt, worked as an apartheid between those who had economic power and the rest of the folk. It was as the powerful people said: *Ok guys, you can strive for these shit places, you can fight for money, political ideologies and racial differences — you can kill each other because it's part of this game — but you can't come near where we are living with our families.*

"These rich people made the population disgusted with where they lived to the point where they preferred to die. My grandad took advantage of it," Antony said to himself.

Frankfurt was an oasis of bankers, bureaucrats, and politicians in high positions. Everything there was futuristic; everybody was educated and intelligent. This city was one of the few places where people could still appreciate some classic architecture and ancient masterpieces. Unfortunately, these were mere preserved ornaments.

The Cessna Citation XX landed softly on a heliport at the Frankfurt Hbf, the biggest and most modern train station in Germany. There, anyone could take a bullet train and go anywhere in Europe, including parts of England, Russia, and North Africa. The station seemed deserted, and Antony knew very well the reason why.

Antony, still confused and annoyed, left the business jet, expecting Mr. Aaron to leave it too, but he just opened a window and shouted out some instructions to Antony from the pilot cabin.

"Go to station three. There is a train waiting for you. Just get in and let it take you," Mr. Aaron still looked angry at his passenger.

"Take me? Take me where?"

"Good luck, Mr. Sims," Mr. Aaron just closed the window and took off in the business jet.

Antony just watched the jet take off, wondering how this situation would end.

It was sunny, and the sky was bright blue. From the heliport, Antony could see the Meno River and a huge green forest surrounding it — one of the beautiful landscapes that still existed on this continent.

She wants me to walk a lot. Why? Antony sought, deep inside his mind, some explanation for this kind of game, however unsuccessfully. Before arriving at the train, he needed to take several escalators and walk across many department stores through this huge station. This station was renovated so many times that it became something like a combination between a NASA space center and a big luxury mall.

"Well, let's take a train… here we go!"

CHAPTER 06

The department stores were deserted, and only a few employees walked along this place, better known as the Frankfurt Central Station. An AI controlled several robots and systems which cleaned and did the maintenance of the whole station. This kind of organization was common in big cities like Frankfurt, Vienna, New York, Chicago, Paris, and London — especially after the fall of the Royal family about a century ago. After several riots and pandemics, these cities managed to decrease the flow of immigrants and replace their hard work with robots who never complained.

At station three, there was a train with its doors open, just waiting for Antony to come in. Right after he got in, the automatic double doors closed behind him. Then the train left the Frankfurt Hbr going toward the train line that crossed many forests and the Meno river.

Antony was walking along, crossing several cabins from the economy class until he got to the VIP classes, equipped with luxury bars and restaurants. *Everything that money was able to buy*, he thought as he looked around.

"Please, Mr. Sims, keep going until you reach the business wagon," the AI's young, friendly female voice explained Antony's trajectory. All large vehicles, like trains, airplanes, and cruise ships, had this system where an AI talked through small speakers hidden inside the walls. All wagons were empty, and Antony knew very well who was in control of everything. *What does she want this time?* This question floated through his thoughts when he got to the empty business wagon — a luxurious and spacious place with big armchairs, meeting desks, bars, and wide screens all around. Everything was high-tech and had white and beige tones.

"Your hostess will arrive in a few minutes, Mr. Sims. I suggest you relax in one of these comfortable armchairs. Would you like something to drink?" The AI's voice just waited for the response.

"No, thanks," Antony snorted and then sat down in the armchair closest to him.

"Excellent, feel free to call me anytime."

The bullet train gained speed, left Frankfurt quickly, and followed a train line alongside the Meno River. Despite its speed, Antony could admire the last beautiful landscapes of Europe and the deep blue river. However, something great was coming up. Something near the train made Antony's eyes widen. Four giant robots, about one hundred feet tall, were working inside the river. They had a humanoid shape, somewhat bent, with long arms and legs, but their chest and head looked like a big tractor cabin. They were probably working on something related to the river's drainage because instead of hands, they had big tools like rock drills and bulldozer shovels. They walked slowly along the river like giants made out of tractor parts, and the water reached their knees.

Since childhood, Antony had been mesmerized by big machines like spaceships and trucks, but robots were his favorites, even though he discovered that this word's meaning was "slave" — which made him a little sad. Antony always saw robots like an extension of the capacities and strengths of humans. *Human beings never settle down within the limits of their bodies, for that reason robots exist,* this is what Antony always thought when he worked on robot projects. However, our culture had infantilized these machines, made them just an inspiration for toys' manufacturers and animation creators. That was the only moment of the trip when Antony smiled a little and felt comfortable. These big machines made him remember amazing moments from his childhood, when he spent all day just playing with his toys and his mini computers, when he looked into how these machines were made in different parts of the world. In his teenage years, when he worked on robot projects in his family's corporation, Antony found, searching on the internet, a community located in the countryside of Kansas, USA, where some guys used parts of airplanes, buses, trucks, and other odd materials to build robots. There was a TV show named *The Maximus Arena* where giant robots built by these guys fought like they were barbarians in ancient Rome. This show had been a great success for six months, but it was canceled suddenly, and nobody knew why. *This show was great... and insane,* Antony whispered to himself when the giant robots made him get lost in his memories... until the AI's voice spoke:

"Mr. Sims, Prime Minister Orlyk is here."

Antony's smile vanished instantly, and discomfort took over his body again.

"Ok... time to play."

The automatic double doors at the end of the wagon opened, and through them, a tall blond woman in her fifties walked confidently in toward Antony — she was Yelena Orlyk. Despite her age, she had an athletic body with wide shoulders, similar to a professional volleyball player, and her black blazer accentuated her superior appearance. Her short, wavy blond hair was impeccable — like she had just left the hairdresser — and her deep green eyes seemed like they never lost their seriousness. She didn't need to speak to prove to the people around her that she controlled everything; her presence alone expressed it very well. Furthermore, she had been the prime minister of Germany for fifteen years.

"Mr. Sims. You don't know the pleasure it is to see you again after such a long time," Yelena sat down in the armchair opposite to her guest.

"Don't bullshit me, please. Get to the point."

"You... Americans, and your *polite speech*. I really appreciate it. But, you know, it's just a friendly conversation," Yelena kept a discreet smile on her serious countenance. Her words came slowly, but her voice was strong and deep.

"You need me more than I need you. You know that very well; after all, you haven't done all this for nothing. So... what's the matter?" Antony's expression and voice were a mix of sarcasm and annoyance.

"Calm down, boy. First, I need to show you something very special, and to do that, we must arrive at a specific place. For now, just relax... Bar, a triple whisky neat for me, please."

The AI's voice said, "My pleasure," then two long, thin robotic arms appeared, moving along through discreet rails on the ceiling of the wagon. Then, they poured a glass of whisky at the bar. Afterwards, one of these robotic arms left the drink on the table in front of Yelena, who took a long gulp.

"Humm... refreshing! Help yourself any time, Mr. Sims."

Antony looked disgusted and impressed with Yelena's gesture.

"Wow... That's why I don't believe in women with Soviet last names."

"It's not Soviet. It's a Ukrainian last name."

"But it was in the past."

"I see you love these ancient fairy tales about how this continent was in the past. Anyway, we have enough time to talk until we arrive at our destination. Well, what about you? What have you been doing apart from your book? I haven't seen you since that last business meeting with your family in... London, I guess."

Antony's disgust turned into rage as he realized that Yelena was more talkative and smiled a lot more than she used to. However, he knew she was trying to unsettle him. She was not a person who said long sentences. This lady only gave orders and never looked anyone in the eyes unless it provided her some advantage. Once a German betting website created a survey: *Do you think Prime Minister Orlyk has teeth?* Ninety eight percent of people answered no. On the following day, the website no longer existed.

"Business? Do you think the damn agreements you made with my relatives were business? You murdered people for your own stupid interests."

"Murdered... no... just ended them. The concept of murder changed thanks to your grandfather. If there is no more death, murder is an ancient concept."

Prime Minister Yelena was also head of the European Peace Ministry, or EPM. Names that indicated an alliance of conspiring powerful members or monopoly had been erased from dictionaries and changed to more friendly names like agreement, meeting, treaty or... *none of them.*

In Yelena's mind, the division of countries only existed because of differences in languages and races. If it depended on her, there wouldn't be any frontiers in Europe, and everybody would speak English. It hadn't happened yet because there were social and political groups that were opposed to it. However, there was something worse... really worse.

Yelena made a terrible agreement with Sims Corp, more specifically with its main representatives, Clifford and Christine, Antony's cousins, and his uncle Peter. The technology created by Allan Sims wasn't able to transfer the consciousness of living people; that could only be done seconds after death or if the person was in a coma. But a genius found a way to transfer people into Cloud while they were still alive.

At first, Antony's plan was to create a program called *Mirror Effect*, that would be able to find criminals and take them to Cloud. This way, this kind of people could be easily watched and judged without the risk of them committing other crimes in reality. It sounded like a good idea, but some of Antony's relatives — who were also his business partners in Sims Corp — had other plans for this system and Yelena was interested. Without Antony knowing, Clifford, Christine, and Uncle Peter sold this system's access to Yelena, giving her immense power over Europe. She could transfer whoever and however many consciousnesses she wanted to the system with just one click on her computer. *Mirror Effect* was a prototype that Antony truly regretted having invented, and thanks to this program, thousands of people had been taken to Cloud by Yelena and her *cronies* from Sims Corp to further their own interests.

These atrocities were committed with no mercy: vacating huge territories inhabited by refugees, eliminating a group of protesters, or *shutting down* political opponents. The list of unethical actions was long, and it was one of the reasons that had made Antony give up his family's business once and for all.

After the Prime Minister justified her murderous practices as if they were just business, Antony became increasingly angry — and started sweating a lot.

"Business? You sent about twenty thousand free citizens directly to a prison in Cloud just to help your real estate developments and agreements over frontiers. Who judged these people? You yourself? You can't image what these prisoners' existence is like... they are in a pathetic limbo... a purgatory with no exit, where their consciousnesses just float through pictures and boring songs. Mindcages were created to detain criminals who are waiting for judgment, not for people who get in the way of your filthy business."

"In other words, they are still alive," Yelena completed Antony's argument. "If Allan Sims was able to create a kind of heaven, the rest of his family was able to create a kind of hell. That's the new circle of life, thanks to your grandfather."

"I could talk all day about your atrocities, but I'm sure nothing can affect you —"

"Affect me? That's for poor people who don't know what they want out of their lives. We are more practical nowadays. Also, I remember this case you talk about. This crowd that was always making protests, didn't pay taxes, and moreover they were living in an area that was rich in valuable ores used to build computer circuits... circuits used by your family's corporation, I suppose."

Antony tried to calm down, but instinctively he knew he couldn't break the iciness of this woman.

"Can you remember where this crowd was from? Ukraine, your beloved land... and you didn't show the slightest bit of mercy before throwing them into a virtual limbo."

"I never liked that land. To be honest, I think this idea of nationality is a waste of time, a stupid reason for people to think they are better than others... Anyway, I would like to talk about something more interesting. Your book, Mr. Sims. I have a lot of questions about it."

Another person who read my book today? Antony thought and his doubts just intensified. Finally, he calmed down a little.

The bullet train maintained its trajectory alongside the Meno River until the tracks turned on the ruins of Russelsheim city, the only part of journey where the landscape was desolated, and got close to the Reno River, one the most important and long rivers in Europe. After the train crossed the ruins of Russelsheim city, the landscape became green again with its dense forest. After a few more miles, suddenly, the train started reducing its speed.

"Well, now everybody has decided to read my book. I don't know why... wait... I know," Antony assumed he knew the matter which Yelena wanted to talk about.

"So, if you know, just talk."

"Ok, but first... what are you even worried about? The population is decreasing more and more, and you and your damn pet bureaucrats are doing nothing to change it. Am I right?"

Yelena realized that Antony was talking about a grave world problem, but that wasn't the matter that had made her take him from Brussels. *He didn't even imagine the real bizarre situation he was about to see.* However, she let him speak, simply because she was enjoying to seeing his nervous temperament and to gaining time too.

"Oh... can you explain it better?"

"You know that better than I do. In the last ten years, nobody has been born in this planet. NOBODY. Can you image how couples around the world feel when they try to have a kid, to build a family, and the woman simply can't pregnant? The population is afraid. Just look at the parks, the malls, kindergartens. All these places are empty —"

"That is not our responsibility. This matter belongs to the Department of Populational Research, and they have affirmed —"

"Fuck what your department is affirming. The truth is that the last kid who was born was a boy called Afonso Sanchez in Salto, a city in northwestern Uruguay. The boy is ten years and eight months old... TEN YEARS OLD, the youngest citizen of this planet. After him, there haven't been any real newborn registries... just fake registries in hospitals and the Baby Registry Services." Antony took a deep breath, "everybody is afraid of some kind of new virus or whatever it is that's stopping new families from being formed..."

"Well, for me, it's just our planet seeking an equilibrium for its number of living beings."

"The media is saying nothing about it. There are just people going to Cloud, but there haven't been any births in the last ten years, just fake edited news with old recordings of babies, creating a damn illusion that there are babies somewhere on this planet. It's clear to me... the world population has decreased about one billion while the population inside Cloud just grows every single day... and you're fucking worried about that? You just talk about that like it was a kind of *planet reaction*. Of course, Cloud is to blame for that, but... damn... I can't prove it yet," Antony looked very disappointed and calmed down completely.

"But you have explained it in your book."

"It's just... just a tale in my book. Nothing real."

"Nothing real?"

Yelena calmly took a phone from the pocket of her blazer and accessed a digital copy of Antony's book. She went directly to a specific page and started analyzing some sentences.

"According to your book, in one of its tales to be more specific, *we are imprisoning our souls here on Earth, preventing them from going to heaven and coming back later into new lives.* Interesting, in my opinion, but there is more. *If we keep them in this golden cage, they can't be reborn.* You were talking about Cloud, I guess."

"I just told to you. It's a tale that I created based... based on my dreams, that's it. You can't use it to explain the destiny of our population," Antony seemed uncomfortable and a little shy.

"But there is more, and this part really drew my attention," Yelena continued: *"One day, winged creatures will come to our world through an immense Black Hole... to destroy this golden cage. These winged creatures, armed with swords and axes, will finally set our souls free from this cursed invention of humanity.* This last one sounds very apocalyptic."

"Wow... I was probably very drunk when I wrote it."

"I confess that I imagined angels instead of a drunk guy writing... Anyway, for me, there is something real in these tales, something you want to get off your chest, but you're still afraid to do it."

"Well, I appreciate you having read my book but... hey... what's that?"

The train had decreased its speed even more when it turned onto the tracks over the Reno River, and something in the air drew Antony's attention. Some old army drones were flying around, making Yelena's guest even more confused.

"Granat model army drones? Actually, they are *flying museum pieces.* I haven't seen these in a long time, probably because... wait —" Antony realized something that made him very worried. "They are controlled by an ancient signal called *radio signal.* So, I can't access them. You are certainly hiding something from me."

"Well, Mr. Sims. I know we have had our disagreements, but from now on, I really need your help," the coldness of Yelena's expression was unbreakable.

The train stopped completely on a bridge in the middle of a wide part of the Reno River, as if the train had parked on a huge extension of the deep blue water, and the sunshine highlighted its sheeny, white shape. Yelena tapped on her phone then got up and stood, waiting.

"As the folk from your country say: *Let the magic happen.*"

The roof and wall in front of them opened as if the wagon were a big toy. In a few seconds, the wagon was completely open, and part of the wall had transformed into a kind of dock on the river, but not as long. Antony got up too and noticed something different high above the river, as if it were printed on the sky. From this moment, Antony's face showed only astonishment, even though he had already seen it in his nightmares.

About one hundred yards above the waters of the Reno River and about two hundred yards from where Antony and Yelena stood, a circle of smoke in shades of black and purple was in the sky. Its size was almost the same as a train wagon, and its smoke circles flowed into itself in continuous movements. Some particles of smoke spun around themselves and vanished as they revolved, at the same time, as a whole it stayed stabilized in the same space.

"It's... it's a Black Hole ... up in the sky... that's more than insane," Antony was completely mesmerized and afraid of what his eyes and his super developed mind couldn't assimilate.

"If it's insane for you, imagine what it is to me and to the people below me. Thirty-two hours ago, this Black Hole appeared there, and luckily, I was the first person to be notified about that," Yelena spoke like a general beside her guest. "There was no surprise when one of my officials told me that he'd read about that in your book. Despite demanding complete secrecy about it, this occurrence has spread out through the scientific community on the deep web. Every scientist has warned another, and every one of them bought your book in the last hours, I presume."

"I wasn't warned about this, and I'm a scientist too —"

"Who lives drunk and alone," Yelena completed Antony's affirmation. "A few hours ago, the search for answers in your book became a race against time. Like I just mentioned, I read your book too, and I was impressed with the tales where you described Black Holes... like that. Your description is exactly like what we are seeing now. Coincidence or not, I didn't have a better idea than bringing you here as soon as possible... and make you see it with your own eyes."

Antony was frozen, keeping his eyes on this freakish vision while the Prime Minister spoke.

"For me, the main question is simple, Mr. Sims: Do you think *they* can come here?"

Yelena looked sternly at Antony, his eyes wide and his mouth open.

"Well, first... I'll accept that whisky, please."

CHAPTER 07

It was Saturday morning. On the heliport of a penthouse in Brussels, a guy was sitting on a beach chair drawing some sketches and drinking beer. Beside him was a small table covered in beer bottles, several sketches, and a high-tech phone. Despite the sunshine seeming intense on the horizon, the temperature wasn't high enough to dry the ground that was wet with dew.

Nivedita came upstairs carrying a squeegee and started cleaning some parts of the heliport.

"Nive! You don't have to clean here. Come on, don't tire yourself out for nothing," Antony called her out.

"I know I don't have to, but it's good exercise, bright boy. You, on the other hand, should practice some physical activities too."

"I have lifted a lot of beer bottles. One day, I hope, it'll strengthen a lot of my arms' muscles," he said, holding a beer bottle.

Antony picked up his phone and checked a live transmission of a phenomenon that had driven him crazy these last few days.

"Six days since it appeared, and I can't explain why it's there."

"You told me you have seen it in your dreams. You wrote about it. You probably know about this strange situation better than anyone."

"But the reality is different. Since I was a child, I have had dreams about things like this. Black Holes, mythical creatures, monsters… I thought when I became an adult all of it would vanish. But no… and for the first time, I'm confused about my own thoughts."

In spite of his jokes, Antony felt deeply frustrated not having found any rational explanations for all that. Doubts, and a sense of weakness made him sit for long hours in silence since he came back from Germany.

A military train adapted for this particular situation stood where he had been six days ago on the Reno River. On this train, drones and cameras monitored the Black Hole twenty-four hours a day. Antony had access to the cameras through his phone, with Yelena's permission — if she hadn't given him permission, Antony would have gotten access his own way.

The news of this strange situation with the Black Hole hadn't spread through the media, but on the deep web, it had become the most researched topic — especially after some conspiracy theorists discovered that the area around the Reno River had been isolated by the armed forces and they hadn't reported the reasons for it.

Nivedita picked up some sketches that had been blown onto floor by the breeze and noticed that they were all the same kind of drawings. Black Holes and bizarre creatures, similar to humans but winged and wearing strange armors. She looked worried as she organized these papers on the table close to Antony.

"Well, my bright boy, I think the answers you're looking for aren't on a website, the deep web, or *anything web*. Extraordinary situations like that require extraordinary attitudes. You know, don't hide who you really are from yourself... you're different, you know that."

Antony stayed silent for a moment. Then he tried to take his mind off his doubts, at least for a moment, keeping his eyes focused on the image of the Black Hole on his phone.

"Once, you told me about Buddha... I mean, Siddartha. You told me something about three or four truths and something about reality. Oh... I've always had a good memory, but right now it's not working so well."

"It's about the Four Noble Truths, and your memory is not so bad; I told you about this a long time ago," Nivedita said affectionately. "Actually, according to Siddartha, humans are subject to desires and cravings, and they waste their lives trying to satisfy them. Therefore, their suffering continues on and on until —"

"Until reality knocks on their door and teaches them that the end of suffering isn't on the outside but inside them," Antony's eyes looked frightened."

"Well like I just said, your memory is not so bad."
"I have... I have to see my mother right now."

Antony's hands were trembling when he took his Headgate from under some papers on the table.

"But you usually visit her on Sundays."

"I changed my mind. I have to go now. Warn me if something happens, please, Nive."

"Do you think a jet might land here today?"

"No... I hope not. That's why I'm up here."

Antony put the Headgate on and fixed the beach chair to lie down more comfortably.

"Ok, Nive, see you soon... here we go!"

"See you... my bright boy."

Nivedita looked up to the sky, feeling like something really bad was coming up. The fading sky of Brussels seemed really calm and silent — which made her feel even more afraid.

CHAPTER 08

A bullet train approached the other that seemed to be a train from the army and stopped parallel to it on the tracks over the Reno River, then a platform connected both. Yelena left the train that had arrived and got in the other one. The one that had already been there had wagons modified to house army technicians and special equipment, forming a kind of HQ. Inside, there were several computers and monitories controlled by German men in army uniforms who observed the Black Hole all the time, twenty-four hours a day. Several small drones equipped with cameras flew around the Black Hole. Also, some miles from there, in a military base, a group of Eurofighter Typhoon Z and jet MIG-39 drones were ready to take action, according to any orders from the HQ set in the train.

The Prime Minister approached the men watching the monitors and surveyed the image of the Black Hole like a lioness looking for her prey.

"Anything new, guys?"

"Some electric emissions and an increase on the speed of smoke undulations. You can check it on this monitor, if you please," a confident soldier explained pointing to the monitor ahead. The screen showed a 3D version of the Black Hole, indicating that the smoke circles were spinning increasingly fast in the last thirty minutes. Moreover, several attempts to enter the Black Hole had been unsuccessful — the drones just went across it like it didn't exist, and only insignificant interferences were detected.

"Keep your eyes open, men. If the tales in Mr. Sims' book are a kind of omen, we shall expect the worst."

***

Some minutes later, the Prime Minister was ready to go back to the bullet train when, suddenly, the one thing they were watching for happened so quickly that it made everyone tense up. Something left the

Black Hole, and almost immediately, it vanished completely in the air, leaving only smoke fragments.

A winged creature, the same height as a tall person, flew away from the train. Nobody could identify exactly what it looked like, except for its wide, black-stained gray wings that looked like they could belong to an enormous eagle.

"LOOK AT THAT... WHAT IS THIS?" One soldier shouted out.

It had been a long time since Yelena hadn't been so mesmerized, but she would never show any loss of control while facing a difficult situation. She commanded the situation as if it were all just one more regular day.

The rest of the soldiers were extremely nervous, trying to locate the creature through army satellites connected to the HQ.

"Give the jet pilots orders to take off. Inform the target's trajectory to them as soon as possible," Yelena was completely focused on the mission.

"Trajectory identified, Prime Minister," another soldier confirmed as he typed quickly on his computer. "The target is going toward... Paris!"

"I don't believe it's going to Paris for shopping... send the target's trajectory to the jets. They have Permission to shoot down the target in case of danger," Yelena ordered harshly.

Two Eurofighter Typhoon Z and three MIG-39 jet drones flew directly to the location given by the HQ, almost at the speed of sound. After a few minutes, they crossed over Luxembourg and caught sight of some kind of creature flying right toward Paris. Fox One and Fox Two were the Eurofighters' codenames, and the jet drones, controlled by pilots stationed in a military base near Frankfurt, responded by the codenames Hank One, Two, and Three. All of them were impressed by the speed of the creature and apprehensive about what they were going to face.

"Attention everybody, Fox Two speaking: Any visual on the target?"

"Hank One speaking: Approaching the target. Focusing camera on it, now."

Hank One's camera got a clear image of the winged creature: A strange woman who had dark beige skin with some black and blue stripes all over. Her body was tall and strong like a professional Olympic athlete or a Renaissance sculpture. She wore a kind of silver armor covering her torso and a short skirt made out of pieces of leather and metal. Her arms were brawny with some metal bracelets around them, and by her waist, there was a long black sheath. On her feet, the jet pilot could see long metal boots and, on her head, long black dreadlocks — some of them red and white.

"Hank One speaking: It's hard to explain... but the target looks like... a woman with large wings."

The winged woman turned her head abruptly, looking directly at the drone piloted by Hank One, as if she was paying attention to every word that the pilot was saying far away from her. This sudden action was so spontaneous that Hank One was startled from in his control cabin.

"DAMN... she's looking at me... and... and she can hear me. I swear."

She had big blue eyes, similar to those of a cat, and thick lips set in a square face with blue and black stripes, like some American indigenous people. Her expression showed a kind of intense wildness, like a beast ready to attack, but patient enough to act at the right time.

"Hank Three speaking: Just calm down and keep the trajectory, guys. I'm gonna get closer to the target."

Only the sound of the jets' supersonic turbines echoed through the air in a clear blue sky as the winged creature flew with her body rigid, as if she was diving fast in a deep sea.

They entered France's air space, as the army train in Germany went back to Frankfurt. Everybody was astonished and anxious, calculating the next move as they watched their target fly free.

The jets and the winged woman were about to cross over the ruins of Reims City toward Paris when Hank Three decided to do a risky maneuver.

"Hank Three speaking: Approaching target to try contact."

"Fox Two speaking: Keep safe. We don't know what it's able to do."

"Hank Three speaking: It's probably safe. The target doesn't seem to hold any dangerous weapons."

The jet drone controlled by pilot Hank Three was quickly approaching the winged woman when, suddenly, she turned her body around completely with agile dexterity and stood in position to attack. The pilot hadn't predicted such a quick movement. He tried to dodge the target to avoid a crash, but it was too late. The winged woman drew a sword, and with a quickly strike, she easily ripped one of the jet drone's wings off. The army aircraft exploded as it fell to one side, and it's ripped-off wing fell to the other, away from it.

In the control room, miles away from the attack, an alert signal started blinking, and Hank Three took his headset off quickly. His heart was beating fast, and his forehead was sweating nonstop.

"Did you see that?" Hank Three looked around to his partners in the control room. "Did you see it?" He said, pointing his sweaty trembling hand at the screen in front of him.

Next to Hank Three's cabin, Hank One and Hank Two were afraid even though they were miles away from Paris.

"Fox One speaking: Missiles ready to take down the target. Just approaching and looking to get the best aim. Fox Two cover me. Do you copy?"

"Fox Two speaking: Covering you. Missiles ready too."

The winged woman glanced back to check where her pursuers were positioned. Then, with subtle wing movements, she started flying in a zigzag motion, making the jets change their strategies.

A huge urban landscape with the Eiffel Tower highlighted right in the middle stood ahead — the view had its beauty, but nothing could take the pilots' focus away from their target.

"Fox One speaking: Fox Two keep giving me cover. Hank One and Two cover the sides. In case I miss the shot, you must shoot to try to corner the target."

All pilots agreed with Fox One immediately, ready to shoot missiles and machine guns.

The pursuit between jets and the winged creature was like a dance of eagles over Paris' sky until Fox One got his aim exactly on the target who seemed to lose speed suddenly.

"Fox One speaking: shooting."

A missile shot toward the woman who stopped, floating in the air, and used one of her wings as a shield, just waiting. Then, the missile hit the target, transforming the woman in a big ball of fire that lit the whole sky. The jets flew away from the explosion and its blinding light. Then, from a cloud of smoke, a human shape in flames, seemingly unconscious, fell rapidly toward the downtown of Paris, leaving behind a long smoke trail through the air.

***

The extensive security team of Paris consisting of cops, firefighters, and armed forces, was on high alert, trying to predict what the point of the city the creature would hit. Firetrucks, ambulances, police cars, and military vehicles raced through all streets and alleys that were accessible with their sirens ringing, making people leave the streets and run for safety in their homes or any nearby establishments.

Some teenagers, frightened by surge of vehicles, ran into a Coffee Shop for safety. One of them pointed to the sky, and showing the others a winged creature in flames, falling like a comet on a huge mall's dome on the next block. The woman crashed through the dome, creating a large hole, and continued falling into the mall — violently hitting the spacious floor amid wide corridors, escalators, artificial trees, shops, and some frightened people. Some of them clutched phones with trembling hands, trying to record the freakish scene in front of them.

All of Paris' security team vehicles parked around the mall, and out of them, the armed ones ran into the mall through all accesses — front doors, garages, and some of them broke stained glasses with battering rams. The movement of the crowd was frantic and intense.

On the floor, inside the mall, the woman in flames got up and stood as if the brutal encounter she had just gone through hadn't hurt her bones and muscles. She made quick, distinct movements with her wings to get rid of the flames on her body while many cops, snipers, soldiers — came around and kept themselves ready to shoot at any moment.

There was a dense cloud of smoke around her body, but little by little, it was dissipating, filling the surrounding area with a kind of gray mist.

Tension, anxiety, fear… all these feelings took over the atmosphere of the place. Everyone was just waiting for any quick movement from their target to get into action. *Did it come from another planet? How did this monster survive? Can it kill everybody here?* These were some questions that floated through the minds of everyone around the creature.

The flames were put out, and the smoke finally dissipated completely, revealing a muscular, strange woman wearing armor that had only a few black stains caused by the fire. Only half of the wing that had been hit by the missile remained, and some gray feathers spread around on the floor still burned. But the woman stayed calm as if she hadn't just lost part of her body.

The clicks of guns being unlocked echoed, and bullets were about to rain on her. But the woman shocked everyone one more time. She stood still and raised her hands in a gesture of peace, as if she had given up her plans. However, she kept her expression serious and brave… really brave.

CHAPTER 09

Inside the church in Cloud, Antony and his mother were kneeling, praying in front the stone sculpture of the book and Pisano's "Virgin and Child." Antony never visited his mother any day other than Sunday, but an extreme urge to go to church made him change his mind. Angela never complained to her son about the fact that the church was never given a name, probably because it didn't matter to her. Both Angela and Nivedita had taught Antony how important it was to have what they called a *religious feeling,* especially these days. Allan Sims' invention transformed death into a product or in a simple process. Religions and their gods became useless as atheist ideologies gained power and influence. Sacred books were burned, and sacred places were destroyed. Only a few churches remained in the world, but they had become just empty libraries or clandestine habitations for the homeless.

Mother and son left the church arm in arm and walked down the quiet streets where only a few children played football, and others rode their bikes and scooters. Smoke coming from several homes' backyards spread the smell of barbecue through the air. Beyond that, most kids were holding ice creams.

Despite the happy environment, Angela realized that her son had said nothing since they left the church. A mysterious silence that wasn't typical of Antony.

"Has anything bad happened, Tony? You're so quiet today."

"Anything bad in me being quiet? You're always saying that I complain too much about this place."

"But I don't say that with the intention of repressing you. I just think it's funny. I always hear you say things like: 'Poor people, eating their fake barbecue, they will never know the pleasure of having a stomachache or getting fat, or look at these children... they play all day long and never sweat.'"

It had been a long time since Antony had smiled; it was a shy smile but it was enough to make his day lighter.

"Ok, ok... you're a specialist in jokes now, I love these observations."

"I learned it from the guy who loves to complain about everything. It would be amazing if you could come here every single day."

"No, I can't. Sorry. I can't stand being in this place every day. Look at these children and... humm... Why is everybody here getting ice cream?"

"Today is Ice Cream Day. The kids love it. While they enjoy this day, their parents enjoy a delicious barbecue. Just smell it in the air."

"Ice Cream Day? There is a Spring Day, a Popcorn Day, a Baseball Day... Why can't there just be a *Day Day*? It's too many things to celebrate."

"That's my boy." Angela looked relieved. "Sometimes you look like your grandfather when you talk like this. I would love it if you could have had the opportunity to meet him."

Apart from the videos of Allan Sims making speeches to thousands of businessmen, pictures with those same men, and interviews on the most important channels of the world, there was nothing else about him, especially about his personal life. A powerful man like him was able to delete anything about himself from the Internet, at least anything he didn't want people to know. Despite Allan having passed away when his grandson was just a baby, Antony never showed much interest in knowing more about him; not even the reason why he didn't send his consciousness to Cloud, his own invention. However, this isn't what Antony was here to talk to his mother about.

After some minutes of having a lovely walk and conversation, they arrived at Angela's home where a breakfast of cakes, cookies, different kinds of bread, tea, and coffee was set on a table on the porch.

"Hey, look at that! A huge breakfast... with no beer. I probably won't survive."

"I love and hate your jokes at the same time." Both took their seats at the table. "Have you talked to Nivedita? I'd love to see her here on a visit, of course."

"You know, she never agreed with the concept of this place. *If someone passes away, let the circle of life take care of it the way it should be*, that's what she always tells me when I ask her to come visit you," Antony mimicked Nivedita's voice.

"Oh... very well, to be honest, I was expecting this to be her answer. Nivedita has always taken her traditions very seriously," Angela looked a little disappointed.

"Just like you, I guess. How about your old friends? You're so different. I have never seen a picture of you together and... to be honest, I've just realized that I have never seen a picture of you when you were young... and I have already peered into a lot of our family's photo albums... even the ones hidden in their phones. You don't even imagine what some of them hide."

Angela looked a little nervous, like she was hiding something, while she poured coffee for herself and her son, but she disguised it discretely with a shy smile.

"Well... you know, I never cared much about taking pictures, and neither did Nivedita. When we're devoted to a religion and start understanding it well, we leave behind some things that just... you know, inflated our egos like taking selfies, going to parties, following fashion trends... anyway."

"Yes, I understand."

"And you know... it's really impolite to invade the other people's private photo albums," Angela spoke in a serious tone but still sounded affectionate. "Do you want sugar?"

"Yes, please... ok, you know I don't like the rest of our family very much, but sometimes I get curious about what they're doing."

"Maybe there is a longing inside you. An urge to join the family again."

"I don't think so… For me, they are so stupid. I only like to talk to you and Nive."

Antony drank a sip of coffee and watched some adults and their kids holding hands, walking down the street. "It should be amazing living with their kids in a world where there is no pain or death," he thought as his face expressed doubt.

"I love talking to you and Nive about religion, even knowing almost everybody on this damn planet talks about it as mythology. I drive Nive crazy when I talk about religions as if they were just fairy tales, but she knows it's just a joke."

"Is there something more you want to know?" Angela's maternal instincts detected some doubt on her son's expression.

"Yes… Nive always tells me about Buddha… I mean Siddartha… And you sometimes tell me about other special men like Jesus, Moses, Mohammed, Joseph from Egypt… these men… you know… had strange visons, mysterious dreams, and when they tried to warn people about something important, they were treated like madmen or as a danger to society… So…"

"So?"

"So… do you think I could be like them? A kind of messiah? Even in a world where religions and gods don't exist anymore?"

Angela smiled shyly as usual and walked closer to her son.

"A messiah! Someone who comes to set a new order based on peace. That's a strong position."

"Yes, sorry. I should be humbler," Antony looked scared.

"No worries, but it's not the point. Did you notice what some of these men have in common in their stories?"

"No, I didn't."

"They had similar stories... like a walk through the desert... imagine these men were walking there, especially at night. Imagine the silence, the loneliness. Just you and the stars as corporation. There, they knew themselves... it's a meditation. A meditation that you face in your desert. It would be amazing if people were born with an instruction manual, but they aren't."

Angela caressed Antony's shoulder as he stayed silent just looking at the folk walking on the street in front of them.

"So... keep crossing this desert, my son, and the answer will come."

"Yes... I see... I'm tired of being alone, but I don't want another option."

Suddenly, Antony heard a signal from outside, then he put his hand on his ear like he was taking a call on a headset. It was Nivedita again, warning him about something very strange but that he had already expected. "Ok, Nive. You can connect me with him," Antony's expression became very tense. He got up quickly and walked in circles on the porch, making his mother nervous.

A few seconds passed like an eternity inside Antony's mind. He hoped that what he'd just heard could be a lie, but after the last incidents in Frankfurt, he believed everything.

"Ok, I'm going right now."

"What's happening?"

"Love you mom... no worries."

Antony kissed his mother's forehead so as not to give her an opportunity to question his actions. Angela closed her eyes when her son kissed her, and when she opened them, he wasn't there anymore.

Only the sad mother remained at the breakfast table, watching families walk happily on the street in front of her home.

***

On the penthouse's heliport, Antony opened his eyes and took his Headgate off quickly as a Cessna Citation XX flew over him.

Mr. Aaron's voice came loudly through the speaker of Antony's phone on the table beside him.

"Mr. Sims, sorry for disturbing your sunbathing but —"

"Ok, ok, spare me your damn jokes. I'll just leave the heliport free for you. Just wait a little."

Antony took his stuff and left it on the corner of the heliport.

"Well, next stop… Paris!"

CHAPTER 10

Antony Sims couldn't control his anxiety as the business jet Cessna Citation XX approached Paris. *How can a creature from my nightmares get to reality?* That was the question that he asked himself nonstop in his mind. His mother just had just talked to him about ancient prophets who had visions and strange dreams. *Where could these visions and dreams come from?* This was another question that annoyed him. If someone from another world wanted to warn us about something but couldn't come here, it would probably try to warn us in other ways: a vision, a dream, voices, premonitions. Throughout history, many people have been called crazy, mad, or heretics for claiming to get visions or hear voices — maybe they had a capacity beyond human limits to hear beings from other worlds, like someone with perfect hearing in a land of deaf people.

Antony looked up to the sky and started imagining humans' attempts to communicate with beings that are invisible to them — from other planets, other spiritual levels, life after death, parallel worlds, voyagers from past or future, the subatomic universe. All of this was fascinating and controversial, but if beings from a different world came to ours, people would see them as a threat — at least that was what fiction movies and books had shown.

Through the jet window, Antony looked down at Paris, a city that was a challenge to any sociologist. Not as modern as Frankfurt but and not as dingy as Brussels. This place was like a *bomb* made out of people, people from Africa, the Middle East, and foreign students. A bomb ready to explode in riots against the government and corporations. The disorganized immigration and the high reputation of its universities made this city a land of cheap labor and students connected with political activism organized by non-government organizations. On one of the streets, Antony noticed luxury shops, cafes, where elegant people show off their fashionable clothes and expensive phones. However, on the next, he saw dirty facades, frightened people, most of them wearing work uniforms, and piles of trash. The only thing these streets had in common was the presence of many cops and private security on every corner, no exceptions. Under the Eiffel Tower, there were many accommodations for immigrants and the homeless, a combination of slums and something like warehouses — one of the results of the violent riots at the end of the 21st century. This Tower had been built to celebrate an ancient revolution in this country, one of the few parts of history still taught in schools — the fact was that this story inspired activists and militants to occupy this area with protests, riots, and battles between civilians excluded by the state and the state army. The civilians won the battle, but they were constantly watched by the state as a *potentially dangerous crowd* — in truth, a union between the state and some powerful corporations made this chaos possible so that they could have effective control over society themselves. From the jet, it was possible to see, inside the big tower, hundreds of beggars living in tents through the corridors, and on its rusty iron structures there, was a lot of colorful graffiti.

The world has become increasingly mad, and something worse is coming, Antony thought.

The jet crossed Paris downtown and flew to the military base far from there, in the commune of Vincennes, located on the east of the French capital. After some minutes, the aircraft finally landed on an area next to Chateau de Vincennes — an ancient medieval fortress with some modern buildings and wooded areas around it. All kinds of military vehicles, including tanks and trucks-mounted howitzer, were parked around there.

Antony Sims got off from the jet and met some soldiers and members of the French security before entering the fortress. Inside, he was searched and submitted to every security protocol. He felt nervous and just agreed with the officials who approached him to provide information about the security protocols for this unusual situation.

In a control room where several computers and soldiers worked with absolute focus, Antony met Prime Minister Yelena Orlyk. From this moment on, they forgot their differences and treated each other like former workmates who were bent on solving a matter of life and death. There was a wide windowpane that separated the control room from a spacious area in the middle of which was a large cage.

"Is that her? She looks like a normal person... ready for a costume party, of course," Antony was mesmerized and creeped out.

"Yes, it's her. Destroyed a droned that costs five hundred thousand euros like a child breaking a toy, and the roof of a mall as well. She only surrendered after her fall, then the military brought her here," Yelena explained assertively.

"Have you tried to make contact?"

"Just the basic: 'What's your name? Where are you from?' But she didn't respond. To be honest, I don't know if she knows our language. She just stays there and looks at everyone... like she is ready to attack."

The winged woman was sitting down on an iron bench like a big bird in a cage. She was quiet with her hands on her knees and a serious expression that never changed. Her wings were half open, and they took almost all the space inside the cage, especially the one that was still whole.

"I want to try to make contact."

"Are you sure? You look afraid," Yelena contested.

"Yes, I'm sure. I have seen her in my nightmares. I have to try and end this agony." Antony seemed increasingly nervous.

"You're brave," Yelena said, then she looked at the soldiers around them. "Place the snipers around the cage and get a bulletproof vest for Mr. Sims."

"This woman destroyed a jet like it was a toy, as you just said. A bulletproof vest won't save my life. I have to show trust. She knows something about us," Antony refuted.

"Yes, you're right, Mr. Sims. In the mall, she held up her hands while dozens of armed men stood around her. It sounds odd, but she knew the meaning of some gestures, I suppose."

"I hope she doesn't know the meaning of being put inside a cage. That's insane but... here we go."

A heavy automatic door opened, and Antony entered an area that looked like the interior of a large warehouse with white and gray walls. On the floors above, several snipers were positioned with their guns ready to shoot the strange prisoner. Antony walked carefully until he was in front of the woman he had only ever seen in his nightmares. The spotlights on the roof lit the tense scene that seemed like an interrogation, and in the control room, everybody was alert, expecting the worst especially Yelena.

Antony stood in front of the cage, a few steps away from the winged woman who gave him a cold stare.

After a few seconds of silence that seemed like hours, he was hit with a strange feeling. An epiphany, an insight, a realization. Just the presence of this mysterious being unsettled Antony's neural network. She made him calm down and remember several nightmares he had had since he was a child — strange names, concepts about life and death, visions of creatures — Antony was sure he had seen this creature before, but in that moment, he hoped his head wouldn't explode like it had in most of his dreams.

Just a few more seconds of silence. Antony's head was still intact, which made him more confident to say the first words that arose from his mind and came out of his mouth spontaneously.

"Muriel. That's your name, isn't it?"

"Antony, my Messiah... I know you have been hearing me all this time."

Everybody in the control room was whispering: *they know each other! How is it possible?* Except Yelena, who just paid attention to both.

Muriel spoke slowly and calmly, displaying a strong confidence and sometimes showing a discreet smile that broke through her serious expression a little. She looked up and around her sometimes as they talked. Her voice was sweet, and everyone got to hear each word clearly — a voice completely different from what would be expected from her warrior appearance.

From that moment, Antony looked a little more confident, though still nervous and confused.

"What are you? An Angel?"

"I am what I am... you know that."

"You look like an Angel to me... so why do you have wings?"

"Angel is just a name created by your ancestors, a name which they gave to my people. For you, it would probably be odd if I asked, 'Why do you not have wings?' I have wings because where I am from these wings are necessary, like fish in your world need flippers or lions need claws. We have what we need to live in our own world." Muriel gave her explanations like a priest in a Sunday morning mass.

"Well... about your world, I mean heaven, I guess... does everybody there speak my language?"

"Heaven is just another name created by your ancestors, and I know many languages spoken in many different parts of the universe."

"Universe? Are you talking about other planets?"

"No... your concept of universe is so primitive, yet... I said world, not planet. There are many worlds connected with each other, especially with your world, which your people call Earth. And each world is dependent on the next one, which depends on the next, building an infinite web of worlds."

"Wow... to me, it sounds odd because... look at us, we are so different, and your world is probably very different from mine. This connection couldn't exist," Antony tried really hard to keep a continuous conversation.

"But it exits because, in the past, we once lived together."

"My people and yours? Together in the past? When?"

Antony had completely forgotten the tense situation he was in and talked to Muriel like he was an investigative journalist, and the Angel was an exotic interviewee, well-behaved on her seat.

"Not just my people and yours, but others too. Let me explain better," Muriel continued calmly. "There was an age when many kinds of people lived together in harmony, right here... on Earth: humans, like you, beings able to fly, to run in the speed of winds, who lived under the seas, giants, dwarves, beings able to do what humans named magic and witchcraft... the list of different beings is immense. Beings that humans, after a long time, named myths, legends, demons, gods, and other names. Beings that humans, through the ages, have eternalized through art in museums and sacred places and by tales, poems, and stories."

"It's insane... and incredible. But it's hard to believe that all of these beings... or people once lived together and then they decided to live on other planets."

"Look at the animals in this world... dogs, whales, eagles, deer, tigers... you are very different, you live in the same world, and you have to respect each other. At the same time, to some degree, these animals depend on the respect of humanity to survive."

"Yes, you're right," Antony was shocked he hadn't realized something so obvious.

"And, like I just said, they are not planets but worlds... worlds that came to be in an age when humans did not have any wish or reason to register their own history in writing. Little by little, each kind of people went to live in their own world, a world which they thought was safer and better to live in... After that, time passed and made humans interpret the existence of these beings, like I just said, like myths, legends, gods, and other names... like they are something far away."

"So far away that no satellite ever found any evidence of their existence," Antony looked a little skeptical. "Well, could I know how you got here? In... my world?"

"Satellites. Human technology is still impressive to everyone in the universe," Muriel smiled, knowing very well that this human technology was useless to watch the places where she had been. "There are people like me, able to travel through these worlds, and there are people like you, who can just listen or see some things from other worlds. You, like some prophets in human history, are different. Do not ask me how you or these other ones can listen to or have visions of beings from other worlds, please. You just are what you are."

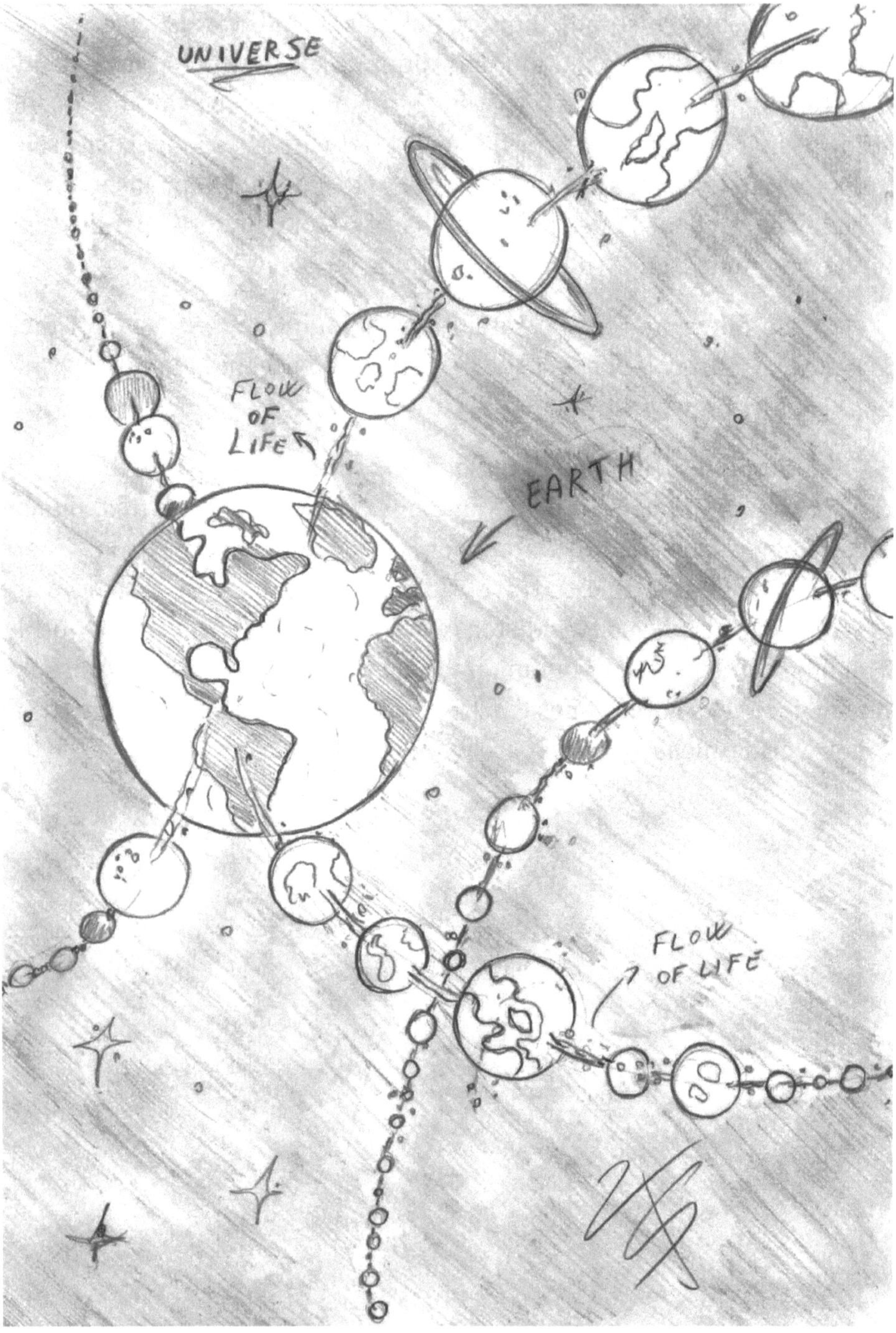
UNIVERSE
FLOW OF LIFE
EARTH
FLOW OF LIFE

"So... you've been trying to warn me about something, all this time... since I was a child, through dreams and visions... Because I'm able to do it. I can still hear you in my mind saying: 'You will understand everything in due time.' What should I understand?"

"As I just told to you... the worlds, despite having been separated, are connected by what we would call a *Flow of Life*. All lives go through these worlds in a natural course, a passage that humans named death... All living beings die one day in one world, then they are reborn in another world where they live their new lives. Then the process repeats itself... they die and are reborn in another world, flowing through the universe."

"Wow... so, can I choose which world I want to go to?"

"What defines the world which you go to after death is how much you evolve... you can go to a better world or you can go worse one. This *Flow of Life* is the only connection between all the worlds and what maintains the existence of the universe... a live universe."

"I could say that the souls or spirits, anyway, are reborn in other worlds."

"You can name that however you want. I prefer to say Life... the essence of every being... the essence which the man whom you called grandfather discovered a way to contain —"

"MY GRANDFATHER? You know my grandfather? How?" Antony's eyes were wide.

"Everybody knows your grandfather where I come from," Muriel continued. "A man who found a way to contain the essence of *Life* inside a technology. A technology named... Cloud. It's an incredible and terrible invention which, unfortunately, worked better than your satellites."

"It's not an essence or anything... It's just memories and —"

"Consciousness... it does not matter what name humans used. The technology created by your grandfather is impeding the *Flow of Life*... keeping the beings from evolving through the universe. And that is the reason why I am here."

In the control room, everyone listened to the conversation downstairs attentively. For the soldiers, this situation sounded like a conversation between two patients in a psychiatric hospital. *This talk is downright crazy!* someone whispered. However, Yelena stayed silent until one of the soldiers asked:

"Isn't it time to put the plan into action?"

"No, It's not. Let them talk more. I have to know what Antony's intentions are yet."

Antony's anxiety was replaced by doubt and fear when Muriel mentioned his grandfather.

"My grandfather... had wings... like you? Did you talk to him?"

"Yes, he had... but now he is in another world far away from where I live."

"Far away from heaven? You mean... beyond heaven?"

"What your people call heaven is just a world connected to Earth and other worlds. Your grandfather, or his *Life*, just continued its course. Unlike humans, my people have more knowledge about the passage through worlds, about the beings who go to live in one world then live in another one and so on..."

"Wow... can you explain the reason a bit better... the reason that brought you here? How is Cloud affecting what you call the *Flow of Life*?"

"Yes, I can, my Messiah. Your grandfather's invention is keeping the lives here from moving on and evolving to other worlds... nobody has been born in my world and in the next one too for a long time... it is causing a complete devastation in the universe. Or like your people say: a *domino effect,* but towards extinction."

Agony took over Antony's body; his chin trembled nonstop, and his face was sweating beyond normal. Also, he realized that the whole enterprise that had been created by his grandfather and managed by his family wasn't destroying only Earth but other worlds as well. He felt faint, and his heart seemed like it could jump out of his chest any time. However, Antony stayed firm in his position, face to face with the mysterious winged woman.

"Oh... so... it's, it's more than insane... what... what must we do?"

"Think of Cloud like a huge dam that is preventing a river from flowing through a dry forest. If humans don't let the water flow down, then the forest will be totally destroyed."

Antony went silent and started shivering.

"My world is dying, my Messiah. Cloud must be destroyed by people from your world, or something worse will happen."

"Worse? What do you mean?"

"The mission is to destroy Cloud and reestablish the equilibrium of the universe. The first plan is just to come here warn people who can do it... If your people don't listen to me, others like me will come to finish my mission... but they will not talk... just act. If your people will not destroy Cloud themselves, my people will do it."

"How much time do we have?"

"Time enough... or another Black Hole will open in another place."

"Where?"

"A place you named... New York City."

"New York City? But Cloud isn't located there," Antony was doubtful.

"My people and I know the best places to spread our messengers. Or do you think I am here because your people captured me?"

"Yes... I... I guess."

"No, my Messiah... I am here because I want to be here."

"What?"

In the control room, Yelena gave the order:

"Contingency plan authorized."

Shortly thereafter, one of the several snipers stationed around the cage shot a tranquilizer dart that hit Antony right on the neck, which scared him a lot. A few seconds later he fell as he cried out:

"Yelena, you son of a bi —"

All soldiers approached the cage quickly, holding their guns. Antony was on the floor, unconscious, and Muriel stayed calm, just sitting down on the chair, watching all that was going on around her. Among the soldiers, Yelena appeared and stood in front of her prisoner. The Prime Minister stared at the winged woman like she had total control over the situation.

"Well, you understand our language and speak relatively well. I hope you listen to me now: If you decide to react, these men around you are authorized to kill you. Do you understand me?"

"I just react when it is necessary... and you are lucky because now it is not necessary," Muriel stated as she stared at Yelena. "Besides, your warriors are not ready for me. You understand me?"

Yelena couldn't show weakness, so she decided to just walk away from the cage without answering her prisoner.

"Men, just follow the plan and take the visitors to their places," Yelena commanded sharply.

Muriel's eyes just followed the Prime Minister as she left the place, and soldiers carried Antony away.

CHAPTER 11

I remember you... you were a child; you came to warn me... You will understand everything in due time. I still remember those words... oh, my head hurts so much; it feels like it's going to explode.

Antony opened his eyes as if he had just escaped from a nightmare, but a nightmare outside of his mind — in real life. Nivedita was sitting on a chair beside him, holding a damp cloth to his forehead. She looked worried. Also, the living room of the penthouse in Brussels was clean, but in that moment, it held two terrible visitors — two bodyguards dressed in the black uniform of the riot police and black helmets. Antony slowly got up and sat on the couch, sore all over.

"Nive, what's... what's going on? Oh... I feel like I've been run over by a tank."

"No worries, my bright boy. Just rest."

"Bright Boy! HAHAHA... I can't believe you still call him that."

This last comment came from the wide screen in front of Antony, where three familiar people were waiting on a video call. The first was Peter, or Uncle Peter, like Antony used to call him. The second was Christine Sims, one of Antony's cousins, as well as the general manager and the main engineer of Cloud Base. Her employees had nicknamed her

The Brain of the Base. And the third was Yelena, who needs no introductions.

Peter was a fat, blond man in his fifties who had a rosy face that constantly showed sarcasm and selfishness. Despite his short, combed hair and executive style, he only got his position as Sims Corp business manager because he was one of Allan Sims' sons. He had all kinds of diplomas and certificates in economics, but he understood just enough about it to pretend he was a hard worker. Unlike Peter, Christine was the most qualified member of the family, with master's degrees in engineering, computer science, and psychology — a cold woman in her thirties who looked a little different from the rest of the family. She was white but had Hispanic traits and long, smooth black hair. Her executive style accentuated the powerful presence of a woman who spoke only when necessary and knew exactly how to get her point across in any argument. However, she had a terrible flaw — Christine always thought she was more intelligent than her youngest cousin.

Antony's face showed a mix of hangover and pain when he noticed these three familiar faces — of which he didn't know whom he hated more. Nivedita stayed by his side the whole time, worried and crestfallen.

"That's unbelievable… you're just missing one more to form *the four knights of the Apocalypse*. What the fuck are you doing here?"

"Very well Mr. Sims," Yelena took the lead. "You gave information enough for us prevent a disaster and prove that you had already had some association with that creature. I don't know how yet, but I will find out."

"WHAT? I have never seen this winged woman before. If you're accusing me, then you have to prove it."

"I saw it with my own eyes. You already knew her name, and she knew yours despite her only calling you 'my Messiah', which, to me, sounded so bizarre. So, what was that? A coincidence?"

"I… I don't know how I knew her name," Antony felt dizzy the whole time. "Her name just showed up in my mind. I heard her name… in my dreams, and when I saw her, I don't know why, but I was sure that that was her name."

"Dreams? Very well, but that's not enough to prove you're innocent of a crime of treason against the state and humanity. I will press charges myself. Thank you very much, Mr. Sims. Now I have to accompany our *winged visitor* to America. Also, stay safe. I may need your help, but I probably won't."

"You said you would help me," Antony's eyes were in flames.

"And you believed me hahahah —"

The Prime Minister laughed sarcastically like a witch from a children's story, a scene that would always be prominent in Antony's mind. Then she ended her transmission, leaving Antony with his relatives.

"Well, that's the first time I ever saw her teeth. Now I'm really scared," Antony looked astonished for a moment and noticed the weather outside.

It was raining, but not very hard and it probably was about midnight. Furthermore, there were two drones flying close to the windows, watching inside the apartment. *One more normal night*, he thought.

"At least she was fast," Antony spoke curtly. "And how about you two. Did you come for a family therapy?"

"Hey Tony, it's really nice to see you again," Peter spoke like it was a happy family reunion. "You have always been the black sheep of the family. Have I told you are my favorite nephew?"

"No."

"Of course, I haven't. I would never say something so stupid. Anyway, Tony, do you see these drones outside? Your only responsibility tonight is to stay where you are right now. Let the adults take care of this situation with the hot-winged woman or whatever."

"Cloud is safe from any interference, thanks to me, of course," Christine spoke in an arrogant tone. "Unlike other family members who just abandoned the business."

"You know you need me to solve problems in this damn Cloud; otherwise, you would have killed me a long time ago," Antony complained.

"I would never kill my own gifted nephew, not only because I can get more money with you alive, but especially because of other things that you hid in Cloud," Uncle Peter stated.

Christine wanted to keep Uncle Peter from talking about what had happened in the last meeting between the three members of the Sims Family and Yelena. But her uncle couldn't control his own mouth.

"I have told you, Uncle Peter, Antony was just bluffing because he didn't like our business with Yelena."

"Can you two refresh my memory? Yelena's dart left a pile driver working inside my head… ahaaaa."

"I'm talking about something that you called Key," Uncle Peter looked a little angry in that moment, "something really stupid that you put in our Cloud."

"Our Cloud? You two always thought you were the owners of the world."

Two years ago, the unethical business between Sims Corp and Yelena — involving the misuse of the program named *Mirror Effect*, created by Antony — made the main heir of Sims Corp so angry that he decided to threaten his own relatives with something that he called Key. Key is a kind of powerful sequence of algorithms that was hidden somewhere in Cloud. With just one command, Key could *turn off* all of Cloud at once, letting all consciousness living there free and ending the Sims Corp business. Christine had scanned each memory, pixel, code, and algorithm in Cloud, and nothing had been found. But Uncle Peter was always suspicious about that. He was not so intelligent, but he knew his nephew was a unique kind of person and crazy enough to do it — and smart enough to hide it even from Christine. Coincidence or not, a few weeks after this business meeting, Angela died of cancer and then her consciousness went to live happily in Cloud. The mystery of Key kept disturbing Uncle Peter, and Antony never said if it was true or just a bluff to scare his relatives.

"*Owners of the world*... sounds good," Peter calmed down. "Anyway, Tony, could you just be a normal playboy? You know, parties, whores, alcohol. Leave this Key story with us and be happy. Look at your mother; she is so happy there."

"It's so disgusting that you base your life only on parties and prostitutes. Look at what you're doing to the world —"

"Let's get to the point," Christine led the conversation, "just stay in your home. These bodyguards will be around while Uncle Peter watches you through the drones outside."

"Yes, directly from my comfortable mansion in California hahaha."

"Moreover, these bodyguards just took all phones and computers from your home, just to make sure you will stay far away from trouble. They'll take your housekeeper to her home too. Please, lady, it's time to go," Christine just observed Nivedita through the screen.

"Can I at least hug my son before I leave?"

"Your son? Hahaha... look at your skin color, lady. You never belonged in this family," Uncle Peter's voice was mocking. "Anyway, just hug your *imaginary son* and get out."

"Don't pay attention to them, Nive," Antony spoke softly. "Everything will be alright soon."

"Yes, I know."

Nivedita hugged Antony and whispered close to his ear:

"My phone is under the bathroom sink."

"That's my girl," Antony whispered back.

CHAPTER 12

Uncle Peter seemed to dance as he walked through the wide and luxurious corridor with a red carpet and many family pictures on the walls. This was just a small part of his mansion in Orange County, California, where he lived and pretended to work from home. Peter walked into a small control room where one single man, wearing a security uniform, watched several screens. This man was easygoing just like Peter, so easygoing, in fact, that he watched the monitors with his feet on one of the keyboards, drinking a cup of hot chocolate.

"Hey Joe," Peter asked as he watched the monitors, "how is our hostage going?"

"Everything is under control, boss. Our hostage doesn't seem to be dangerous."

On the two wide screens above them, they watched images from the drones that flew over Antony's penthouse. The images from outside the apartment were clear and showed Peter and his employee a guy who looked a bit drunk, relaxing on a couch, watching TV, and drinking beers. There were also a lot of beer bottles on the coffee table. It was a clear night in Brussels, no wind or rain, according to the images on the screens.

"I don't believe that this is the most intelligent guy in the world," Joe looked disgusted. "Looking at him, I don't feel so bad."

"Very well, but keep your eyes open. This guy loves to mess up and make surprises."

"Look at that... he is putting his hand in his pants. He's watching porn, I guess."

"Oh, damn... I have already seen enough for one night. Anyway, I'm going back to my office. Let me know if anything happens."

"Ok, boss."

CHAPTER 13

In Brussels, the rain came down harder, and a guy wearing a blue nylon jacket held an old phone in his hand and ran up the fire escape to the heliport. The people close to Antony always had stories of the *miracles* and exploits he could do using all kinds of computers or phones. *Just give him a cellphone with only one button, and he can control anything;* that's the kind of sentence Antony's former mates used to describe him.

After finding Nivedita's phone hidden under the bathroom sink, Antony quickly planned his escape and a way to meet Muriel. *What kind of mad experiments could Yelena or the American government be doing to her?* He couldn't stop thinking about it. Antony was sure that this woman he had seen only in his nightmares was just the tip of an immense iceberg, something bigger that could affect the whole world, and not even his family or the government could prevent it. *They manage everything as if it were a business, and not even a Black Hole and a winged woman can prove to them that something supernatural is happening;* this was another thought that annoyed Antony.

On the dark horizon, something was approaching, gradually decreasing its speed. It's a Bombardier Global 1K, a business jet, not as modern as the Cessna Citation XX. However, this one was specially modified to attend the needs of Sims Corp, and these modifications made this aircraft the most modern in the world. It was the first business jet equipped with an engine that used chemical elements that produced nuclear energy — not enough to cause a catastrophic nuclear explosion but perfect to keep this aircraft flying for a long time in high velocity. Also, this business jet had powerful retractable turbines like the Cessna Citation XX, which enabled it to land without a long airstrip.

Antony stood on his penthouse apartment's heliport, just waiting for the business jet to land. The Sims Family had gained innumerable resources from their investors, sponsors, and even from governments around the world, they were like kids with so many toys that it was impossible to play with all of them through just one childhood. The family got most of these goods as bribery in return for the best places in Cloud. Summing it all up, the Sims Family owned mansions in several countries, apartments and vehicles, modern cars, jets, speedboats, and motorcycles, and Antony could control any of them through his phone.

The Bombardier Global 1K landed and opened its door automatically, then Antony quickly got in. This Jet was dark blue with white strips around it and a white infinity symbol on its fin that seemed to shine.

"Welcome, young Mr. Sims," the AI's voice said when Antony entered the jet.

"Young Mr. Sims? Only you call me this, hahaha... Ok, let's follow the route I sent you."

"Impossible, young Mr. Sims, the system is not authorized to follow this route."

"No problems, I'm gonna download the system for you in a few seconds. Just take off. Be quick."

"Ok, young Mr. Sims."

All business jets were equipped with a specific system that enable them to fly in different conditions — short routes, through storms, or even in cases of air battle. In this case, the Bombardier Global 1K had only been authorized to fly in Europe, but Antony just downloaded a system that enabled the jet to fly for longer routes outside the continent. Inside the jet, Antony had all he needed: laptops, phones, drinks, meals, a comfortable armchair that turned into a bed, and another very special vehicle.

"Ok, Jet. Next stop... New York."

"Ok, young Mr. Sims."

CHAPTER 14

Uncle Peter was at the office in his mansion just checking some documents on his computer when suddenly his phone rang.

"Hello!"

After a few minutes.

"WHAT? But I've been watching him the whole time... oh, damn... let me check this out."

Uncle Peter's face was red and furious as he walked quickly to the control room. There, he found his employee, Joe, in the same position he had seen him before, drinking coffee, his feet on a desk. Peter's furious entrance made Joe jump from his chair.

"I TOLD YOU TO LET ME KNOW IF ANYTHING HAPPENED!"

"Hey boss... calm down, please," Joe was trembling. "Your nephew is still in the same place. Nothing changed since you left."

Uncle Peter saw in the monitors that Antony was in the same position that he had seen previously — drunk on a couch, watching TV. From Brussels, one of his henchmen warned him that a big Sims Corp business jet had landed on the penthouse and quickly taken off.

Peter kept watching the monitors, trying to understand what was happening. He soon noticed something odd — and suddenly he realized. In the images shown on the monitors it wasn't raining, but the information from his henchman in Brussels told him that it was raining a lot at the moment.

"DAMN IT... it's just a recording."

When Antony left Sims Corp two years ago, many gossip TV shows started following him around to find something controversial about him, but they were all unsuccessful. Once, one of these TV shows decided to stalk him with drones, which just ended up recording a drunk guy masturbating and watching porn. A few weeks later, this kind of TV show stopped pursuing him because most of the audience was disgusted with this harassment. But Antony still remembered these videos, so he downloaded them from internet on Nivedita's phone, then he edited and sent these videos to the drones' system. Just Antony's genius could hack the internet and program systems this way, so fast and with just a simple phone. In short, Joe had been watching a loop of old videos. It gave Antony enough time to call one of the family's jets and escape.

***

Uncle Peter went back to his office, and from a safe behind a bookshelf, he took a big green tattered suitcase. It was an old piece of military equipment from the 20th century, a type of telephone that used a radio signal. The Sims Family kept two of these, one in Uncle Peter's office and another in Christine's office in the Cloud Base, located in the middle of the Nebraska desert. This kind of communication was used by the two when they wanted to talk about Antony — because any meeting through internet could easily be hacked by him. Peter looked indignant as he called his niece to inform her of the situation. The communication through this equipment was very bad and full of interference, but it was the only way to keep a conversation from Antony.

"Hello, Chris… our little genius is probably going to America. He has stolen one of our best jets."

"I thought this could happen. It was a mistake letting you take care of this situation," Christine always talked calmly.

"Hey, take it easy. I'm not as smart as you. And I kept him away from any computers."

"It doesn't matter anymore. He is on his way, so let me take care of this my way when the time comes."

"Oh, you... you wouldn't dare to reveal the secret to him, would you? This can hurt the boy a lot. If you do this, he can use Key just as revenge."

"Now you feel sorry about him, hmmm... Come on, Uncle Peter, do you still believe this Key story?"

"Wait, wait, he doesn't deserve this."

"I'm gonna tell him at the right moment. If you don't like my idea, try to stop him before I do this. Ok? I have to go."

Christine ended her transmission, leaving Uncle Peter in silence, and with a bad feeling about his own family.

CHAPTER 15

The dark blue Bombardier Global 1K flew over the middle of the Atlantic Ocean. It was hard to try to define what time it was in that moment, but the sky was as blue as the business jet, making it almost invisible. Antony was lying down on a bed that earlier had been just an armchair. He watched the deep orange horizon through the window and was mesmerized by what the planet could offer — a beautiful landscape that had never been touched by humans, as like if the planet had just told everybody: *Hey, you... you don't need to change me... just enjoy me*. This thought came to Antony's mind and made him want to read a part of his book. He took a tablet from a compartment next to him and opened his book.

***

The Golden Cage:

Under a cloud, two angels, each one holding a sword, were ready to attack a village formed by huts, right below them. One of the angels was a woman, and the other was a teenager. One was the mother and the other was her daughter.

"What is the plan, Mom?"

"Look at the middle of the village, my sweetheart, look at that huge golden box. That's where they keep their souls after death. And these souls are suffering a lot in there. I can hear their screams."

"Oh, Mom! Why have they done this?"

"They fear death, my sweetheart, and they are so egocentric that they do not want their God to take care of their souls. Moreover, nobody is born here... not anymore, not after they built this golden box."

"Oh, it's so horrible, mother. Why does it happen?

"My sweetheart, life and death are sides of the same coin. Humans must die for other humans to live. That is the flow of the universe. Preventing death is preventing life too."

"So, we should destroy the Golden Box, shouldn't we?"

"Yes, we should... but let's wait a little longer."

"Can they see us from there, mother?"

"No, and we can't go there, not yet. I will open a Black Hole in the sky, then we can go there."

The mother made a circular motion with her sword and opened a Black Hole. Then the two angels went down to the village to destroy the golden box.

The villagers, realizing the angels' threat, tried to protect the golden box by throwing hammers and sickles at the winged beings, who just protected themselves.

"Mom, they won't let us approach the golden box. What should we do?"

"There is no other option, my sweetheart. We must kill all of them."

The angels flew nimbly through the villagers and killed each one of them until not a single one was left alive, forming a huge lake of blood. The large golden box remained intact in the middle of all the blood, and some damaged huts floated around it.

The mother angel landed on one of the floating huts and stayed there, crouched, watching the golden box.

Desperate screams came from inside the golden box, begging to be free, while the mother angel stayed still, just watching.

The daughter landed at her mother's side and asked, "Mom, don't we destroy the golden box?"

The mother suddenly looked confused as she gazed at the cracks that appeared on the box, until she answered, "No, I can't... We can't. I realized we can't destroy it. If we destroy it, other humans will come and build another one. Listen, they think they are... they are happy. I thought these were screams of anguish, but these are screams of happiness... I... I don't understand."

***

Antony awoke from a deep sleep with his tablet in his hands, not sure when his reading had become a dream. This was just one of many tales in Antony's book, *Cemetery of Souls*. Many descriptions of the Black Hole, the angels, and the angels' actions were identical to what was happening these days. The idea of being a messiah to warn the world of something important didn't leave his mind, but at the same time, it sounded so egocentric to Antony. *If you have power enough to make the world less bad, just do it* — it was his mother's quote, which his mind insisted on repeating. Also, from his mother's teachings, another quote came to Antony's mind: *Great powers can make a better world... or a worse one.*

Despite the orange horizon seeming to be far away from the jet, like an infinite flaming line crossing a dark blue immensity, it didn't lose its beauty.

"In two hours, we'll arrive in New York, young Mr. Sims," the jet's system AI's voice said, and Antony stayed silent, mesmerized by the view.

CHAPTER 16

One of Air Pizza's drones flew over the suburbs far away from downtown New York, in a region where there were abandoned houses and poorly kept football fields. It was approaching an old, abandoned hangar with an airstrip full of cracks and surrounded by bushes.

Air Pizza had been bought by All Star Food four years prior, after it specialized in the delivery of pizzas and drinks by drones. All Star Food was a company that had monopolized all delivery services around the world and sixty percent of the world's fast-food production. They had decided to keep the name Air Pizza so it didn't sound like a monopoly. However, everyone knew its unethical business practices to control the worldwide food trade.

The drone slowly got inside the hangar, approaching a guy who was sitting down on a wooden box next to an old wooden table. The drone landed softly on the table, then the guy took out his big pizza and a little crate of beers. After some commands on the guy's phone, the drone took off and went back to the closest Air Pizza store. This was the most practical kind of delivery, not only because of its efficiency, but also because there was a lack of young employees to do this kind of service.

The companies that used this system knew that the global population was decreasing, and these kinds of services hid this dangerous problem with technological innovations.

Antony Sims was eating a big slice of pizza fiercely when he noticed three special visitors inside the hangar — three visitors who had sneaked in from under the Bombardier Global 1K. Three different-colored moggie cats appeared and stood just watching him eat and waiting for some leftovers. Antony threw a few pieces to his visitors who ate as fiercely as him.

"Well, you are my only friends these six days in New York. At least you would never try to betray me or make trouble for me, I guess."

One of the cats answered with a meow.

"Humm... I take that as a *yes*."

Antony would like to visit his mother, but he realized, checking the Cloud system, that Christine and Uncle Peter had visited her every single day of the last week. They were probably looking for him, but Antony had been avoiding to meeting them for a couple of years, and they knew it. At least there were cats on the hangar whom he could talk to.

A phone started ringing low on the wooden table, and Antony checked something on it, something important and strange that was happening in downtown New York. Something that Antony had been expecting.

"Wow... time to go, my friends. But there's still time for one more slice of pizza.

Another cat answered with a meow, again.

CHAPTER 17

Breaking news throughout the United States was showing the same event on TV and the internet at around 1:00 p.m. A large Black Hole appeared in the sky of New York City, like a zeppelin hovering over downtown but without moving. On all terraces and windows of buildings, crowds of people holding their phones with trembling hands, trying to take the best picture of the anomaly above them. Conspiracy theories, messages about the end of the world, memes, and fake news were widely spread, even to people living in Cloud.

Actually, all the information about the Black Hole in the media was fake because only one group of people knew the truth, and they weren't too worried about disclosing what this anomaly meant. This group was formed by Yelena's entourage, some members of the American government, and the military — American and European.

This group set up an improvised HQ at the top of an old skyscraper downtown. This floor had high ceilings and walls designed in a neoclassic style with wide windowpanes. The HQ was separated in two parts. The first had a wide control room, where European and American soldiers worked, including Yelena. This part was protected by a wall made out of bulletproof glass, able to stand the impact of missiles.

On the other side of the wall, there was a big iron cage where Muriel sat on an iron chair — calm, with a serene expression on her face. The wing that had been hit by a missile was whole again because of her supernatural capacity to heal and restore parts of her body. At that moment, Muriel was ready to attack, but as a great warrior from her world, she knew the exact moment to act. Beside the cage was an iron table with a box made out of bulletproof glass, and inside it was Muriel's sword, like a rare diamond in a museum.

"I hate her. I would love to kill her. But I can't yet," Yelena said as she stared at Muriel from the other side of the bulletproof glass wall.

"Murder solves nothing. You know that. Especially nowadays when no human is ever completely dead like it used to be in the old days," a man in a wheelchair spoke beside her.

"She is not human, and killing her may send her back to her world," Yelena answered without turning her eyes toward the prisoner.

The man in the wheelchair was Richard Sowell, an ex-military man in his forties who had become paraplegic after a car crash the same year he started a campaign to be elected the next president of the United States of America. Despite this terrible accident, he won the elections. *A strong body that depends on a wheelchair*; that was how everyone saw the President, fragile but powerful. A white man who had stern green eyes that observed everything through round glasses. His short curly hair was always carefully styled, and nobody ever saw him in public dressed in anything other than a gray suit and red tie.

There had been several rumors about his accident, claiming that it had been a terrorist attack because he was a favorite according to the election polls, but his strong body got through this tragic event, and he never admitted the possibility of this supposed attack. The way he talked was often calm and confident, always getting directly to the point — the attitude of someone who knew the power of his position in his country.

"Have you had news about that boy Antony? You told me that thanks to him you discovered where the next black hole would appear," Richard asked Yelena as he looked at Muriel.

"Mr. Sims helped a little, but he is not mature enough and isn't mentally prepared to deal with this problem."

"Helped a little? In my opinion, his predictions about where the Black Hole would appear were conclusive. Look up at the sky! Everyone in my country and inside Cloud is afraid. I have to give a statement to the population about it as soon as possible."

"Don't worry, Mr. President, we know exactly what we have to do."

A voice coming from the computers' speakers echoed through the control room.

"No, you do not." Muriel said through microphones installed in the cage. "If you left your fear of losing power aside, you would know the right action to take."

"I'm keeping you here," Yelena looked a little angry, "it doesn't matter what comes out of this Black Hole. I'll kill it, and then I'll kill you. And I hope the next creature doesn't take as long as you did to show up."

"I told you. I am here because here is the place where I must be. But I admit... this cage is too strong for me to scape by myself," Muriel looked up to the sky through the iron bars and windows. "And about the other of my kind, he is coming now and... he likes to act, he doesn't talk as much as me."

"Ok, Yelena," Richard spoke calmly, "I let you use the biggest city in my country as a trap, now take precautions. I'm going to my improvised office, try to calm my people down on social media and keep talking to the military. Excuse me, Prime Minister."

Yelena just watched President Richard Sowell go to a little office adapted for him on the same floor with several advisors. She could give orders in Europe, but she knew the limits of her power in America, the only part of the planet where she didn't have that much influence. The election of a man who held ideologies completely contrary to hers was a hard blow to her plans. This president was completely against the services provided by the Sims Family and Cloud, and this daring choice in the last elections proved that a huge percentage of people agreed with him. A religious man who still held on to his beliefs was risking his position and the safety of New York City by bringing Muriel here. But a fascination to see this woman and the feeling it was his duty to manage this situation made him take the lead in this mission. Moreover, Muriel told them where another Black Hole would appear — a place where everyone could see and be made aware of the great changes that were to come.

CHAPTER 18

A constant high-pitched hum came from inside the Bombardier Global 1K as the three curious moggie cats watched a side door open slowly. From the jet's door, an iron ramp arose, and the buzz got even louder. Lights flashed from inside the dark cabin where a silhouette of a person riding some type of machine could be seen. The buzz of the accelerator increased, and Antony got out of there riding a motorcycle, a Kawasaki Ninja H5R — the perfect combination of Sims Corp technology with Japanese mechanics.

Allan Sims was a lover of high-speed vehicles, and because of that, he had bought many Japanese factories years ago, just to test his inventions. In the Kawasaki factory, he invested and innovated in engines powered by chemical elements that produced nuclear energy, like the jet, that were able to provide high performance to motorcycles but were also a big danger in case of accidents. Antony inherited his love of high-speed vehicles from his grandfather. Actually, Antony loved sports and their challenges, but because of his heart problems, he was advised to stop enjoying the pleasure of taking part in them. However, he had compensated with the adrenaline that high speed provided him. When he still had a good relationship with his family and work, he could barely wait to have free time to go to the Kawasaki factories and test some new motorcycles for hours, especially this one that he kept in his aircraft.

The motorcycle ridden by Antony was the most modern and fast one in the world, like everything that belonged to the Sims Family. This fantastic vehicle was dark blue, like the color of the jet, and he wore a dark red helmet with white stripes and a green visor. This helmet was connected with the on-board computer of the motorcycle, which had a voice control system and showed him all necessary information on the visor through an image projector. A modern phone coupled with the on-board computer completed all the high-tech equipment, but there was one thing that didn't seem to fit in with the rest — a simple red analog wristwatch that looked like it had been bought at a gift shop.

"Goodbye, my friends, see you soon," Antony said his goodbyes to the cats while the jet door closed automatically at his back. He could still hear a meow coming from far away.

The Kawasaki Ninja H5R got speed easily through the roads toward downtown New York. The roads were empty for a while, and around them, there were just abandoned parks and houses, like the ghost towns in apocalyptic movies. *If it looks sad now, it'll get even worse,* Antony thought.

He was approaching downtown and could see the skyscrapers even a few miles away from arriving. Above the skyscrapers, on a blue, clear sky, a Black Hole in tones of black and purple was still in the same place — from Antony's point of view, it looked like a big purple dot that stood out in the sky.

Some military aircraft and helicopters were flying over downtown, waiting for a worse situation than what happened in Paris. Antony was still trying to imagine how Yelena had convinced the President and the military to set a trap in the middle of New York. Despite knowing where the Black Hole would appear, the staff at the HQ thought Muriel could attract the creature who was about to come.

Racing through the road, Antony remembered the ending of the tale that he had been reading on the jet on his way to America.

***

The Golden Cage:

The mother angel and her daughter were still standing on one of the huts floating on the lake of blood. The mother looked very nervous hearing the screams of happiness from inside the golden box while her daughter said anxiously:

"Mother, we must destroy the box. It's our mission."

"We can't, my sweetheart. The only ones who can destroy it are the humans themselves. If we destroy it, they'll learn nothing and they'll build more and more golden boxes.

"And what should we do, mother?"

"Wait... only the humans can destroy this box... I mean, cage. At least one of them... just one of them has to destroy it, and the idea will spread through everyone else's minds, and they will never want to build another golden cage again."

"But... there aren't any more humans! Look around, there is only a huge lake of blood, no humans."

"No... there are humans... inside... they must destroy the golden cage. It's easy. Look at the cracks! Just one would be able to destroy this horrible prison."

"No, mother, it's our mission. We must finish it. If you won't do it, I will."

The young angel flew and landed on the golden box. She was preparing a heavy strike of her sword when suddenly her mother drew her own sword with one hand, stopping the strike.

The young angel gave her mother a stony look and said:

"Mother… it's our mission. We must finish it."

The mother said in a serious tone:

"We came here to do the right thing, not to do what we want or what we assume to be right."

With one fast, strong strike of her sword, the angel mother tore her daughter in two. The blood of the young angel splattered over the golden box, and the screams of the people inside it never stopped. But after that, cracks started to appear all over the surface of the golden box.

"Yes, humans. Keep going. I know you can. Unfortunately, you only change when you smell blood."

The angel mother stayed on the golden box, watching more cracks form between the two parts of her daughter.

***

Antony had always seen these stories in his mind as fairy tales, but now they sounded like prophecies — maybe there was something hidden in them that could explain what had been happening these days. He knew deep inside that something dangerous was about to happen, and only Muriel could prevent it. However, if the others, who kept her caged, insisted on seeing her as an enemy and not as a savior, the situation would probably get worse… much worse.

CHAPTER 19

As the Kawasaki Ninja H5R approached downtown, Antony realized many huge homeless camps all around. All kinds of ethnicities could be seen there: Latinos, Asians, black people, white people, but the majority were Americans. An immense crowd forgotten by the authorities, who were just waiting for these people to exchange their miserable lives for a life in Cloud, but that wasn't happening. Above them, there were train tracks and roads that connected the rich suburbs and cities to downtown New York, the place that had become a kind of island with beautiful buildings in the middle of an *ocean of homeless people*.

Nobody drove through these roads except for social welfare employees and cops. So, a few people looked suspiciously at the guy riding an expensive motorcycle on a dusty road. Most of them lived around downtown, and probably, after death, they would live in the worst neighborhoods of Cloud. Despite the social welfare secretly having encouraged poor people to commit suicide to go directly to Cloud, they preferred to live there than to become just a consciousness in a sea of algorithms. The authorities didn't understand it yet — maybe an instinct inside the forgotten folk's mind made them prefer to cling to reality instead of living in a virtual world. This situation only happened in America because the local government didn't have the same courage as Yelena, who shut down hundreds of people who, to her, were just inconveniences.

Antony noticed that just ahead of him, there was massive security wall that surrounded all of downtown and that the train tracks and the roads went through. Since the poor people and the illegal immigrants reached a large and out-of-control number, the richest parts of every city in America adopted the wall as protection from the situation that their own government had created. After pressing some buttons on the on-board computer, Antony located a gate to which only authorized people had access to. So, he easily used his gifts to get an authorization and make the gate open for him. The security guard had never seen someone pass so quickly through the gate, but he thought: *if he has authorization to get in, he is probably someone important.*

On the other side of the wall, the landscape was completely different: a perfect combination of modern and classic well-maintained buildings, clean streets, and wooded parks created a pleasing atmosphere where the air was the purest. *My old New York,* Antony thought — the place where he had been born but hadn't enjoyed the way he liked because of the crazy routine imposed by his family.

Some cop cars and military vehicles drove quickly through the streets toward the same direction. Antony didn't need to search on his computer to know that they were going to the place where the strange situation was happening. Thus, he followed them.

While Antony went to a still unknown destiny, he noticed everyone on the streets, parks, and at cafés was focused on their phones, certainly searching and waiting for information about the anomaly that had appeared a few hours ago above their city. *THE END IS COMING, THE END IS COMING,* a crazy man holding a poster screamed as he crossed the street without looking, causing Antony and some cars to hit their brakes abruptly. "Maybe he's right," Antony whispered to himself, then he gained speed again through the busy streets.

This kind of turmoil often happened during the Super Bowl finals or the ancient festivity called Christmas. Meanwhile, the main news websites were asking people to stay safe in their homes, but it was obvious that it wasn't working.

After a few minutes, Antony arrived in front of an imposing skyscraper called the Liberty Tower, the place that in the past had been called the Trump Tower and had belonged to a former president of the same name. If the change in name was a tribute or an offense to that former president, nobody knew to this day. There were several military vehicles and tanks around the tower, which made Antony sure of Muriel's location.

It's here, he thought as he found, through his on-board computer, military secret signals monitoring the Black Hole over the city. All the information he needed was shown on his helmet visor.

Antony parked his motorcycle near a café, in a spot where he could observe a crew of military men and police officers, ready to attack, even though they didn't know what could be about to attack them.

After scanning all computers and security systems inside the Liberty Tower, Antony started planning how to meet Muriel again.

"Let me see," he whispered. "About two thousand soldiers and cops in and around the building. Muriel inside a fortified cage, and... hmmm... Yelena and the President Sowell together. They are on the last floor. Well, in a different situation, I'd say: It'll be fun. But it's not a tale from a children's book."

CHAPTER 20

Richard Sowell sat in front of his computer in a small office, recording videos in which he asked the people to remain safe in their homes while the military solved the Black Hole problem.

Yelena and a few soldiers kept their focus on Muriel, who was still sitting quietly, as if she was in a meditative state.

"Prime Minister, the Black Hole is showing some instability," one of the soldiers in front of a monitor said.

Yelena observed the Black Hole on the monitor and its smoke traces spinning faster. "I have a bad a felling," she whispered when suddenly a large unidentified object came out of it — something so fast that it left only a lilac trail across the sky as the Black Hole slowly vanished.

"What is that?" Yelena asked.

The men who had been watching the monitors were checking all the possible data with the help of several satellites that monitored the whole country.

Richard Sowell appeared, asking for information and trying to calm his men down.

"Prime Minister, Mr. President," another soldier at one of the monitors said, "one of our satellites detected the object that just left the Black Hole."

"Don't waste any time. Give the order for some jets and drones to intercept its trajectory. Immediately," Yelena commanded sharply.

"But... sorry, Prime Minister. It's flying almost at the speed of sound. I think it's impossible to intercept it."

"Can you check its trajectory?" Richard Sowell looked a little nervous as he asked.

"According to our first readings, it's going straight to... Michigan."

"No, Michigan is just in the middle of the way," Richard Sowell intervened, "it's going to Nebraska. To be more exact, to Cloud."

The President Sowell looked at Muriel as he spoke; then she looked at him and calmly confirmed his assertion.

"Yes, you are right."

"So, what should we do to stop it?" The President asked Muriel as he approached the bulletproof glass wall.

"She's not here to advise us," Yelena cautioned, "she's a trap and my hostage."

"So, let me know what's your plan B, Yelena. Because there is a large unidentified object flying over my country."

"Mr. President, Cloud Base is the most secure place on Earth. If a creature like that," Yelena pointed to Muriel, "decided to attack it, it won't have any choice other than to die."

Muriel calmly interrupted the conversation between the two leaders.

"Sorry, Leader Yelena. I will call you this way. First, we do not die, like you and your people. Second, he has a name. His name is Abaddon."

CHAPTER 21

Between the states of Wisconsin and Illinois, four MQ-15 Reaper drones flew over the area, just waiting for visual contact with the unidentified flying object. All of them were armed and ready to attack. They were agile and lethal, piloted by experienced military men and covered in resistant armors. However, the UFO was something completely different from anything they had ever seen. The only visual register they had got on their cameras was something that looked like a big, winged man who was flying so fast that only images of lilac stains in the sky were captured.

But what seemed to be impossible happened. The four powerful, armed drones were destroyed all at once. The UFO didn't seem to have missiles or any firearms, but all that was left of the drones were pieces in flames, falling from the sky as if these machines had been made out of paper. Then it kept going fast on its trajectory toward Nebraska.

***

The news about the four destroyed drones hit the improvised HQ in Liberty Tower like a punch that bewildered everyone. There was no other alternative except reinforcing Cloud Base's security and waiting for the worst.

"Yelena," the President Sowell approached her carefully, "I believe it would be better to study the possibility of asking our visitor what we can do in this situation."

"She is one of them, ready to attack us along with her partner. Let Christine Sims at Cloud Base take care of the other, visitor and —"

Suddenly, a high-pitched noise came from all computers and phones in the control room. Everyone looked more and more lost with their hands on their heads. The noise was unbearable.

"WHAT IS HAPPENING?"

In the middle of this madness, one of the soldiers was contacted on his walkie-talkie by a security officer outside of the skyscraper, then he passed his hand-held radio to Yelena.

"Yelena, listening."

"Sorry Prime Minister," the soldier's voice was loud and shaky, "something strange is happening right now with our vehicles."

"Can you explain better?" Yelena could hear car alarms going off through the walkie-talkie.

"Well, all vehicles' alarms have been triggered at the same time. Also, all of them have been locked, and nobody here can drive or control them. It sounds crazy but —"

"But?"

"But it seems that someone is controlling all our systems."

"Oh… damn it!"

The assumption that someone was controlling the machines reminded Yelena of one person. A person who made her feel disgust. The only person able to control anything connected with circuits or the internet.

The high-pitched noise stopped with no apparent reason, making everyone stay on edge.

Suddenly, the elevator's double doors in the back of the control room opened, and Antony Sims came out slowly, keeping his hands in his nylon coat's pockets and looking defiant.

"Did anyone here miss me?"

All soldiers were pointing their pistons and machine guns at Antony as he walked toward the Prime Minister and the President.

"Take it easy, guys! All your guns are locked. I have always disagreed with this idea of connecting all guns with an integrated system of computers. It takes away your freedom to act. Don't you agree?"

All soldiers tried, without success, to unlock their weapons as Antony got closer to Yelena and President Sowell.

"Well, I think it's better to talk with no guns."

"I don't think so, Mr. Sims," Yelena gave an order to one of the soldiers next to Antony with just a quick movement of her eyes.

The soldier hit Antony on the back of his neck with the back of his machine gun. He fell abruptly on the floor, then the same soldier stepped violently on his neck, and pinning him to the ground. When the soldier was preparing another strike against Antony with his gun, President Sowell quickly intervened, ordering his soldier to protect the already immobilized guy. An American soldier pressed a blade to the neck of the European soldier who was ready to hit Antony again — it made him freeze at once. All American soldiers dropped their guns and drew their knives, each ready to attack the European soldiers who were holding their locked machine guns as bats, ready to fight. Antony, from the ground, followed the soldiers' movements as they were ready to start some kind of medieval battle. A scene frozen by the silence and tension was about to *explode* in violence. Fortunately, Richard Sowell intervened again with a stern voice.

"Yelena, I must remind you you're in my country, and Antony Sims is an American citizen."

"We don't have time for diplomacy —"

"A foreigner who assaults an American citizen in front of the president can cause serious diplomatic problems. Am I being clear, Yelena?"

"You are threatening my men with knives. How do you —"

"No, Yelena. I'm threatening you. Again, you are assaulting an American citizen in front of the highest authority of the United States. My men are highly trained to defend any American, including the president... and if necessary, to eliminate all threats."

"And you should study more about the American soldiers, Prime Minister," Antony said from the ground. "They always have hidden knives in case their guns don't work."

"Especially after the law that demands guns to be controlled by an internet system was approved by WPU," the President Sowell completed.

The Prime Minister sometimes forgot that she wasn't in Europe, and the President still had a sense of patriotism, unlike her who treated everyone the same way regardless of where they were from. Also, she knew her limits in this country, and she was still conscious that she needed the President's help.

"Ok, Mr. President. Men, let Mr. Sims go," Yelena was incapable of saying sorry.

All soldiers relaxed, and the one who was holding Antony to the ground helped him get up.

"That's why I love my president. Ahh... my neck," Antony said, feeling a lot of pain.

"About the guns, Mr. Sims, can you —"

Before Richard Sowell completed his sentence, Antony held his phone and with one tap unlocked all guns.

"Well done, Mr. President."

"Thank you."

Yelena quickly grabbed Antony's phone, an action that startled him a little.

"Hey, you could ask politely."

"Please, Mr. Sims," Yelena's eyes were piercing, "stop joking and try at least to help us."

"I helped you in Paris and I woke up in Brussels sore all over. Give me one good reason to help you."

"Actually, you have to give us a reason, Mr. Sims," Richard Sowell intervened. "You wouldn't cause all of that just to stay here doing nothing, I presume."

"Sorry, Mr. President," Antony looked a little confused, still feeling his neck and trying to say the right words. "All I want to say is... you must let Muriel free for her to protect Cloud. That's it."

"Oh... You're kidding me," Yelena was indignant, "this beast destroyed a military aircraft and caused big trouble in Paris. And you want us to let it go?"

"Yelena, Antony, please," the President cut in, "whatever came out of the Black Hole is arriving in Iowa, so Mr. Sims... letting this creature free is really unsafe. I prefer you'd talk to her and try to find out what their plans are. Can you do that, Mr. Sims?"

"Yes... yes, I can. But there is only one plan. Destroy Cloud. That's their goal and it will be hard to stop it."

"So, it's time to act," Yelena held Antony's phone. "I'll keep this with me. No more trouble now, Mr. Sims. Men, search him, see if he has another phone or something."

Two soldiers searched Antony roughly. Yelena's order bothered him.

"Hey, what's this? A gift from a poor girlfriend," one of the soldiers who was inspecting him pointed to Antony's old wristwatch.

"It's from my mother. Now, can you open this damned *big aquarium* for me to go, please?" Antony looked a little angry.

After the soldiers found nothing with Antony, they led him to the other side of the bulletproof glass wall. As the glass automatic double doors opened, Antony saw Muriel sitting down on an iron chair inside the cage, in the same position as when she had first met Antony in Paris. Her sword was inside a bulletproof glass box close by.

Antony's phone was left on a table close to one of the monitors, and Yelena decided to keep the glass doors open.

He looked less nervous than when they last met. Something inside him said: *it wouldn't be necessary for me to be here now if everyone decided to practice good deeds.* But reality had been spoiled by humanity itself, and it felt that changing everything was its responsibility.

He stared at Muriel, who felt his presence and stared back at him.

"Well, next time we should choose a better place to meet." Antony said.

Muriel looked confused with Antony's statement, which made him suddenly more serious.

"Sorry, I forgot you don't understand our... sense of humor. Anyway... can you tell me who came out of the other Black Hole? Is it someone like you?"

"Abaddon... yes, he is like me, but stronger and bigger... the best warrior I have ever seen." Muriel always spoke slowly, and her face showed only peace and seriousness.

"Is he going to Cloud? Is he going there to destroy it?"

"Yes, this is our mission. But he has other ways to do it, unlike me... he will not stop until it is over."

"So... how can we stop him?"

"You cannot stop him."

"No? That's it?"

"Yes, it is. Like a fish cannot stop a lion. Like an eagle cannot stop a whale. This is the natural relation between my people and your people."

"And you? Can't you stop him? Abaddon?"

"Maybe... I can try to stop him, and I have great chances of winning."

"Ok, nice... so you can leave this cage. Go to Cloud and stop him," Antony looked increasingly nervous.

"Well... First, this cage is too resistant for me to scape. Second, my mission here is to destroy Cloud; if I try to protect Cloud, it would be just to give you and your people more time to think and decide to, finally, destroy Cloud yourselves. Third, I will do it if you want it enough to free me."

"Want it enough? What do you mean?"

"Mean? You humans are interesting," Muriel's voice sounded more severe. "You created thousands and thousands of laws and theories based on what you call rationality, and you preferred to exchange your gods for them. On top of that, you don't follow your hearts, don't defend what is good. If you just followed the goodness, you would not need to imagine what you call heaven or create one... you would already be living there."

The Prime Minister looked indignant, belittling the conversation between Antony and Muriel. "Oh, no. This crazy talk again," she whispered.

Antony was frozen and weeping when he looked at the people on the other side of the bulletproof glass wall.

"I... I understand," Antony said, "but you are strong. You can escape from here easily, I guess. You... you were hit by a missile; you're strong enough to do anything."

"Maybe this cage can be destroyed by me, but to do it, I need proof."

"Proof? What proof?"

"Proof that humans want to change this world... the changes must come from within you. I am just a way to put humanity's will into practice."

"Within us!"

"Yes, escaping from this cage is easy... but escaping from the cage inside our minds — a cage where our willpower is confined — is extremely harder."

Antony felt the immense weight of responsibility on his shoulders. His heart beat fast, and an intense flow of several plans and predictions crossed his mind.

In the control room, Yelena ordered the soldiers to be prepared to act.

"Keep your eyes open, men. She is trying to deceive this poor guy in order to escape."

"She doesn't seem dangerous, I suppose," Richard Sowell said, "but our time is short."

"Really short, Mr. President, I'm gonna ask him to come back. Fortunately, his phone is here."

Antony took a deep breath as tears fell on his face. *It's not a fairytale; it's your life and it must be written the best way possible*, he thought when, at the same time, he pressed a button on the side of his wristwatch.

The phone close to Yelena started emitting an unbearable loud buzz that made everyone put their hands over their ears again. Also, the glass door closed quickly, leaving just Muriel and Antony on the other side. Antony had programmed a kind of trap on his phone connected to his simple wristwatch — in case someone took his phone again like what happened in Brussels. Also, this trap was still full of surprises. The big cage opened automatically, like a big iron flower blooming, and the glass box where the sword was kept did the same.

"SOMEONE, OPEN THE DOOR," Yelena shouted furiously.

Some soldiers tried frantically to open the door through their computers, but it seemed useless. Others pointed their machine guns to the wall, but all guns were locked again.

On the other side of the glass wall, Muriel got up, gorgeous and brave, with her gray wings completely open as Antony grabbed her sword inside the glass box with his two hands, because he didn't imagine how heavy it was, and he almost fell holding it.

"Wow... this is damn heavy. How can you handle it?"

Antony used all his strength to throw the sword that Muriel handled easily, as if it were a plastic toy, then she put it in its sheath.

"It is a matter of centuries of practice. Now I must go, Antony. Bless your god for your attitude."

Muriel positioned herself toward the wide window, like she was a runner ready to break a record. The feathers of her open wings contracted; her feet were fixed on the floor where some cracks started showing around her. With a single strong movement of her body and wings, Muriel leaped through the window that broke into hundreds of pieces. Moreover, her wings moved the air around mightily, pushing Antony back and making him fall again. He lay unconscious on the ground full of pieces of glass, and Muriel vanished through a blue and clear horizon, getting speed easily.

In the control room, a soldier hit Antony's phone with his gun, making the unbearable noise stop. The others finally got the door open through the computer system.

President Sowell rolled his wheelchair closer to Antony, who was still on the ground, trying to wake up.

"Sorry, Mr. Sims, but due to the circumstances, I have to order your arrest."

"Ok... ahh... Mr. President. To be honest... I deserve it."

CHAPTER 22

In the middle of the Nebraska desert, an imposing hexagon-shaped building, about eight hundred square feet wide and one hundred feet tall, stood out in the landscape. Viewed from above, this place resembled a gigantic white heart situated in the midst of a barren landscape, surrounded by a network of roads and train tracks that looked like arteries extending in all directions. There were hundreds of wide numbered doors made out of thick iron all around the building. These doors could only be unlocked after a rigorous set of security protocols. Numerous army cannons and missile launchers were stationed both on and around the building, ready to defend what lay inside the impressive engineering masterpiece. This place was the *Cloud Base*.

Inside this heavily fortified structure, there were countless corridors, walls made out of carbon fiber, and miles of fiber optic cables, enough to go around the Earth more than five thousand times. In the *Master Memory,* a nickname for the place where the majority of the computers which kept people's consciousness and memories were set, separated into many sectors — a complex and infinite number of aluminum boxes that looked like refrigerators with blue and red lights flashing nonstop. The layout consisted of expansive corridors and galleries, spread through six floors, like the sumptuous Egyptian galleries of pharaohs, but made out of metal and twisted cables — all of it powered by atomic generators situated on a high-security floor underground.

Sometimes, an employee of Cloud Base would check on some of the computers in Master Memory, but most of the work was done on the top floors of the building, where about one million robots and humans worked together in long corridors. There were hundreds of desks with numerous monitors — creating a typical sci-fi movie setting in shades of white and gray with hundreds of high-definition screens. On another part of the last floor, there was a gigantic control room called *Sector Three,* where thousands of people dressed in military uniforms worked diligently in front of their monitors, maintaining the Cloud Base's security — overseeing the drones that flew outside, the system of army cannons and missile launchers, robot guards. In short, this sector was responsible for overseeing all aspects of security.

On the other side of the last floor, there was a sector that looked more like an immense hotel, or how most employees called it, *The Human Sector.* There, there were meeting rooms, executive offices, restaurants, and several luxurious suites. This part of Cloud Base was designed for when businessmen and politicians came over. There were other accommodations for employees who worked there for short periods of time; then, they were relocated. Working from home wasn't an option; all the work was done in Cloud Base, except for the members of the Sims Family.

Also on this floor, there was a discrete but no less important place — a place where everything was decided. A huge office at the end of a corridor where there were three thick, wide iron doors that were kept open and programmed to close only in case of danger. This place was Christine's office — the heir of Allan Sims and the main person responsible for everything regarding Cloud Base and inside it. She had a complete view of each part of Cloud Base from two wide screens on her big desk that had several touch screens. Some Headgates were set on the desk, ready to be used and take anyone into Cloud.

Christine was on her desk, focused, checking every possibility and probability for her next moves against the creature who was coming to Nebraska, straight to Cloud Base. Behind her were long glass shelves with several strategy books, awards, and degrees — just an extension of her narcissism. Everything in this office was high-tech, clean, gray, and without expression like Christine or, as some employees gossiped, a *colorless person*.

She was facing this situation with the winged creatures from the Black Hole like a simple security protocol that had been disobeyed, or a garbage bag that should be left in front of the house. She believed a lot in her own capacity — and it could be a terrible mistake.

"Sector Three, any anomalies? The UFO can arrive anytime," Christine asked through the screens above her.

"No, Ms. Sims. We're just waiting," a male voice answered from the screen.

"Ok, keep me updated about any changes, please. And —"

A big rumble echoed through the wall, and another one next. Christine looked up, trying to control her fear because she knew that all walls in Cloud Base were massively thick, and for something to make this kind of noise, it was probably incredibly strong.

CHAPTER 23

"Sector Three, do you copy?"

"Yes, Sector Six… something exploded in the Master Memory. The flames are spreading to another sector."

"Ok, the fire suppression system is working, but the flames are still intense."

"Sector Three, this is Official Alex from Sector Eight speaking. Do you copy?"

"Yes, Official Alex."

"Check the energy system of the Master Memory in Sector Eleven, please. Something happened with the power transformer in this sector. I'm gonna check it out with some men. All armed."

"Ok, Official Alex. Sector Eleven has no power, but it is already being powered by Sector Ten, that's still showing stability for now."

"Sector Three, Alex speaking. Getting to Sector Eleven. Flames and smoke are obstructing our vision ... OH DAMN... THIS IS CRAZY... OPEN FIRE."

"Sector Three speaking, Official Alex, what's happening?"

"This is fucking big... and he is holding a sword... oh, damn... my men are —"

"Sector Three speaking, Official Alex, do you copy? Official Alex? ... oh no."

***

Angela Sims was walking leisurely, carrying a fruit basket in her hand, down a street near her home when she noticed some adults and kids frozen, just looking up at the sky. It was sunny, and the sky was clear except for this instability in Cloud.

All cities in Cloud were created based on algorithms collected through social medias for decades. There were all kinds of cities: metropolises with tall buildings, villages on the beach, and farms — and beautiful suburbs like the one where Angela lived. All places in Cloud were a pure expression of what everybody had wished for in life — places with no violence or crime, where everything was clean and organized. But in that moment, something was happening to the sky in most of these cities.

Angela started to walk more slowly, looking up at the sky as well. She noticed some red and green rays flashing through the sky, like a screen showing interference in the signal. Everyone in Cloud was witnessing the same phenomenon, and it was making them all feel very scared.

A little girl next to Angela started crying as she looked up at the sky. Angela was sure that something dangerous was happening outside, and there was nothing the people inside Cloud could do. She crouched close to the kid, trying to calm her down, speaking affectionately.

"Hey, sweetheart, don't be afraid. This is nothing. Soon, these lights will vanish."

"My mother said that the world is ending," the little girl cried more.

"No, she was mistaken... look, my son is a scientist. A very intelligent scientist. He works out there with a big computer. He has been protecting everyone here, and he would never let this world end."

"Do you promise?"

"Yes, I promise. Of course, I promise."

Angela hugged the kid who hugged her back, but Antony's mother kept her eyes up on the sky, just hoping her son was well.

CHAPTER 24

About one hundred soldiers, all armed with machine guns, got into Sector Eleven where the flames were almost under control. They wore black uniforms, bulletproof vests, and black helmets. Sector Eleven was a wide corridor from which you could see the three floors of corridors above, where the soldiers were positioned and spread out, prepared to shoot the creature that was slowly walking down the central corridor downstairs.

Suddenly, among the explosions and clanging of metal, Abaddon appeared. A big and muscular man who was eight feet tall, holding a long metal sword. His skin had a lilac tinge and bulging veins mixed with white stripes all over his body, like a tiger. His white, short hair was spiky and seemed to shine in the flames, and his eyes, which were entirely white, glowed with the fire. Two little white horns on his forehead completed the odd combination of an angelical appearance with monstrous features. The serene expression on Abaddon's young face contrasted wildly with the destruction he was causing with calculated movements of his massive body and silver wings. His broad chest was bare and full of scars — which showed that as a warrior he had faced, without fear, hundreds of battles. He was wearing only some kind of leather shorts and a big metal belt from which hung several long gray leather strips and chains, forming a kind of long dress that went all the way down to his long gray leather boots. He wielded a lengthy sword with a broad, serrated blade and had two large metal boomerangs with sharp edges affixed to his metal belt, alongside a lengthy metal sheath. On his wrists, he had heavy metal cuffs — which made him look like a medieval barbarian but looking like a hero from Greek mythology.

The soldiers stationed upstairs couldn't believe their eyes. Then a deep voice coming from Sector Three gave the order:

"FIRE."

All at once, they opened fire on Abaddon, covering his body with thousands of fiery points and smoke. Numerous spent cartridges clattered through the corridors and staircases, the metallic clinks echoing against the floors. The clatter of machine guns added to the already deafening noise that reverberated across numerous sectors of Cloud Base. *Maybe this is what hell sounds like*, some soldiers thought.

Bullets rained on Abaddon for two minutes. Then, the soldiers ceased fire and waited for the smoke to dissipate, expecting to see the invader in pieces. However, what they actually saw was so unexpected it froze them in place. Abaddon stood in the same place he had been before, looking as if not a single bullet had reached him, as cold and serious as a statue. Only a few bullet marks remained on his body — no blood, no broken arms, or broken legs. Abaddon's face was still serene, like a marble sculpture. The soldiers were astonished, paralyzed by the circumstances and unable to take any action. Some of them desperately began to change their bullet magazines.

Abaddon sheathed his sword and swiftly drew the two metal boomerangs, hurling them towards the soldiers stationed upstairs. The boomerangs spun rapidly, slicing through any soldiers in their path with deadly precision — the bodies looked like they had exploded, such was the force of the blades' velocity. Desperate screams echoed through the place like a phantasmagoric symphony. No matter how agile the soldiers were, the blades tore through every single one of them. The boomerangs ricocheted off the walls and the metal columns, maintaining their trajectory and tearing through the other soldiers as if they were made out of paper, gushing blood throughout every corridor. Sometimes the boomerangs made their way back to Abaddon's hands, and he quickly threw them again, continuing the carnage that made Sector Eleven look like a slasher movie scenario. Blood poured out from corridors upstairs, and in a few seconds, it looked like some kind of waterfall, carrying pieces of flesh and blood-soaked black fabric.

Abaddon killed one hundred men just standing there, effortlessly, with only a few movements, without ever changing his calm demeanor. This great warrior started walking slowly down a corridor next to Sector Eleven, until he noticed a massive iron door with *High Voltage* written above it and several warning signs around it. Brandishing his sword, he approached the door and drove the blade in between the sliding doors. Every employee of Cloud Base felt helpless as they watched, through the surveillance cameras, the movements of this great creature. Something within Abaddon detected the smell of fear and worry from the crowd in the building. *This is important to them, so I must destroy it,* his instincts confirmed to his reason.

The great winged warrior used his strong hands to open the heavy doors completely, then he entered another vast area containing large power reactors.

The biggest reactor at the end of this sector looked like a huge fluorescent blue light bulb, about thirty feet high and twelve feet wide. Many cables connected this reactor to other smaller reactors upstairs, an important piece of the electrical engineering that distributed power to the computers in Cloud. If that were destroyed, other reactors in Cloud would be automatically redirected to redistribute the power, which could cause a dangerous short circuit and, consequently, shut down many sectors. One single strike of Abaddon's sword on this reactor could bring about an absolute disaster.

Everyone in Cloud Base, including Christine Sims, felt desperate, their nervous systems completely out of control. *It's here, I can feel their fear intensify,* Abaddon thought, and the atmosphere filled with the people's dread confirmed it.

The muscular warrior was approaching the bright reactor, wielding his sword, when suddenly someone came whirling from above and stood right in front of him, ready to attack. Now, a winged woman was standing in the way, keeping Abaddon from destroying the reactor.

"My Muriel," he said, "did you escape from the humans?"

"No, my Abaddon. The humans gave me a chance to defend them."

Abaddon spoke slowly in a deep voice that echoed terrifyingly through the sector.

Both warriors respected each other and didn't show any anger, despite Abaddon having had suspicions about Muriel at first. But that didn't last long.

"The humans gave you a chance, or your young Messiah let you go?"

"My Messiah set me free, and he is a human. So, the humans gave me a chance to protect them."

"I understand... and this is great. But why fight the inevitable? Let me destroy this place that is killing our people and end this pain."

"If the humans set me free, if the humans trust me, I believe they can change, so... I must defend them. This is the wish of our God."

"Yes, you are right... let us fight then, and our God will decide the victor."

"Right."

Muriel and Abaddon began to fiercely and rapidly exchange powerful sword strikes. Abaddon's towering height of eight feet was in stark contrast to Muriel's five feet nine inches, but that didn't mean he had an advantage over her. Muriel was faster and jumped on Abaddon with her wings wide open. On the other hand, Abaddon was the strongest, immovable like a rock, and spent more time keeping his defense than attacking. The clashing of metal blades was deafening — it interfered with the flow of power in Cloud, making the whole system unstable.

Inside Cloud, people were desperate, walking on the streets while lights up in the sky blinked more intensely. Angela was hugging the little girl while they both sat on her porch. The little girl was crying a lot, and Angela was very worried.

Surveillance cameras kept transmitting footage of the two creatures in fierce battle near the reactor. Christine Sims, from her office, tried to give orders to send more soldiers to this sector. But almost all of the men were already dead, and the ones who remained were afraid to go.

Muriel hit Abaddon with her sword until his cuffs started to crack. Nonetheless, he seized an opportunity to deliver a powerful punch to her face. For a few seconds, the winged woman stood stunned, and this was enough for Abaddon to strike her right shoulder and rip out the arm with which she had been holding her sword. Muriel was in a lot of pain when she tried to leap toward her sword on the floor, near her ripped arm, but it was too late. Abaddon took one of the heavy chains from his metal belt and used it to lasso Muriel by her neck as she was almost reaching her sword.

The warrior was pulling Muriel closer to him while he mercilessly struck her with his big sword. She didn't have any other alternative except closing her wings in front of her like a shield to protect herself. Muriel's right shoulder was gushing with blood, and with her only hand, she held the chain. However, Abaddon was much too strong and pulled her easily. Each strike of Abaddon's sword ripped out more pieces of Muriel's wings that were falling apart and becoming smaller. Soon enough, the wings wouldn't work as a shield anymore. When Muriel's wings became smaller than her arm, enough to protect only her head, she quickly took a dagger from her belt and nimbly stabbed through a crack on the cuff on the hand with which Abaddon held his sword. The stab was so precise that Abaddon felt intense pain. As his wrist gushed copious amounts of blood, Muriel kept the dagger in place, using all her strength, until he dropped his sword. Then, she quickly spun her body around, getting the sword from the floor and, at the same time, kicking Abaddon's chin. The warrior felt it, and Muriel, who was crouched, struck one of his thighs fiercely. While he fell, feeling a lot of pain on his leg, the winged woman got up and started striking her adversary violently, as he tried to protect himself with his cracked cuffs.

Even though she had just one arm in that moment, Muriel attacked bravely and looked as if she didn't feel pain or weakness. Her wings had been cut to pieces, and the only part of her armor that remained — the part that protected her stomach — fell to the ground during her frantic attack moves.

Muriel struck one of Abaddon's shoulders with all her strength, but he prevented a deeper cut by holding the blade with one hand while he punched Muriel's stomach with the other. The punch was so strong that Muriel was lifted off the ground a little and lost her breath. Abaddon took this opportunity to grab his sword again and to grasp Muriel tight by her dreadlocks. Then he threw her furiously against the floor, as if she were an old rag doll, without letting go of her dreadlocks. Abaddon could hear the bones of his opponent breaking, but still, he instinctively threw her to the other side, dropping her near her sword.

Muriel was disoriented, with almost no strength left when she saw her sword near her ripped arm. Desperately, she slowly crawled toward her sword as Abaddon patiently walked in her direction, holding his sword in one hand and a heavy chain in the other.

Despite Muriel's whole body being injured, trembling, and bloody, her expression was unshakeable.

The muscular warrior lassoed her neck with a chain again and pulled his adversary up slowly, but she still managed to catch her sword with her fingertips in time. Then Abaddon stepped on one of Muriel's feet, breaking its bones and spilling blood on the floor, as if he had stepped on a rotten tomato. He kept pulling Muriel up by the chain around her neck while keeping her foot smashed on the floor, stretching her body and hearing her vertebrae crack one by one. Abaddon was behind Muriel, and it was impossible for her to strike him with only one trembling arm. She tried spinning her body to strike her opponent, but weakness was starting to take over her whole body.

Like a fisherman admiring his catch hanging off a fishing rod, Abaddon stretched the arm with which he held his big sword, preparing the final strike against his opponent. He took a second and said his last words as Muriel got hold of her sword.

"Right... Our god chose the victor. Now our god allows us to end this battle."

"Yes, but you always let your heart beat too fast when you're close to me," Muriel whispered.

"WHAT," Abaddon's eyes widened like he just realized something very important that he shouldn't have forgotten.

Abaddon's heartbeat directed Muriel to where she should make her last strike. She took a deep breath and stretched her arm as far as it would go, aiming her sword's blade right above her chest. She thrust the blade through her own body with a swift diagonal movement of the sword. The blade went through her back and hit Abaddon right in the middle of his chest, straight through his heart. Abaddon froze at once, feeling his heart stop and his strength vanish. The last strike made Abaddon's heart explode inside his chest with a gush of blood that soaked Muriel through.

The two bodies fell to the ground with a heavy thud on a lake of blood, but only one was alive. Muriel, the winner of this terrible and mortal battle, kept breathing without showing any pain or suffering, peaceful and serene just like when she had started fighting.

***

Two soldiers went into Sector Eleven carefully, with their machine guns ready to shoot, but that was not necessary. The two men were completely dumbfounded by the scene that looked like a huge mausoleum made out of metal, flesh, and blood. In the center of it all, two creatures lay torn apart on the blood-red floor. One of the soldiers was scared when he realized that the creature that looked to be the most destroyed one was still breathing.

"That's terrible... ask the infirmary people to come over immediately. One of them is still alive."

"But... it's not human... it seems —"

"That doesn't matter right now. It's a living being. Follow the order, please."

The soldier complied quickly while the other one, who had given the order and was probably in a superior position, said to himself:

"Well... if the Sims Family had created a heaven, they have just created a hell here."

CHAPTER 25

In a gray room with only a simple metal chair and table, a guy sat, handcuffed. His face showed something between worry and indignation until the room's door opened, and an elegant man in a wheelchair went in and positioned himself in front of him. His handcuffs were chained to the table, and he had no other option except keeping his arms stretched on the table and listening to the man on the wheelchair.

"So... Mr. President, did you come here to determine my verdict?"

"No, I didn't, Mr. Sims. I came here to thank you. The mess you made in Liberty Tower was necessary to save Cloud."

"Wow," Antony looked scared and happy at the same time. "So... Muriel probably easily killed the other winged creature and saved the day. Am I right?"

"I wouldn't say easily. The other winged creature was much bigger than Muriel, and he destroyed five floors made out of concrete and iron. Moreover, he almost destroyed one of the reactors. But, thanks to your friend, nothing worse happened."

"That's insane," Antony looked intrigued. "I imagined it would be someone like Muriel, but... my mother? I have to see my mother —"

"Your mother is ok, no worries, boy," Richard Sowell eased Antony's sudden concern. "Cloud is stable again. And your friend, Muriel, has been slowly recovering in the infirmary of Cloud Base these last two days. Her body was almost destroyed, but these creatures have incredible healing power and resistance."

"Two days... wow... I don't have any notion of time in here. What about the other creature? He's dead, isn't he?"

"Well, that's why I'm here to ask for your help again, Mr. Sims. After your friend, Muriel —"

"She isn't my friend... I guess."

"However, you asked for her help, and she complied. Let me continue. Muriel killed the big one, and his body was kept in a laboratory in Cloud Base, but this morning... it wasn't there anymore."

"He's not there anymore? How?" Antony looked suspicious.

"The surveillance system only showed him waking up and leaving, destroying two massive iron doors on the way. Since then, nobody has seen him, and we are afraid this creature may come back to finish his... you know... his mission."

"Oh... I have no idea... even less how I could help you."

"Well, you already helped me once. You can probably help me again. But, Mr. Sims, you told me you thought it could have been another creature like Muriel. You were almost right. And you were sure that Muriel would help," Richard Sowell seemed to interrogate Antony, but without pressure.

"What's your point, Mr. President?"

"Well, I'm intrigued about how you made your decisions. You seem to know more about this weird situation, more than anyone else and more than you yourself are aware of, it seems. Can you tell me how that's possible?"

Antony looked doubtful, and he had no words to express his actions, but he tried:

"So... These ideas just came to my mind, and I... I have always talked to my stepmother Nivedita about it... wow, I miss her so much... anyway, and she has always told me about... our senses."

"Senses? Like sight, smell, touch, etc..."

"Yes... we are limited to these five senses. But what if there are more senses that enable us to see or feel other worlds. What if some people have it, people like me and others throughout history, like Jesus, Joseph of Egypt, Buddha, or Joan of Arc and... well... I don't know, Mr. President, most of these people were seen as crazy, rebels, or ill. You know, not even they were sure about what they could see, but they saw. And nobody could guide them because nobody had this kind of sense they had. For me, they were the most evolved beings of their times."

Richard Sowell stayed silent, and Antony looked downcast, realizing that his speech seemed to be useless to the president.

"Well, I don't think that's enough to make my sentence shorter."

"You belong to the most powerful family in the world, Mr. Sims. It's impossible to arrest you, even if I wanted to," Richard smiled friendly, "but what you did saved billions of people."

"Billions of people? They are just memories, neuro connections floating around in a fake life."

"However, you know, according to WPU, since 2154 the people who live in Cloud have been considered human beings, and any attack against the people who live there will be judged as it would be here, in real life, as I call it. So, you saved millions of lives, and according to WPU you're a hero."

"Yes, I know about the Twentieth Treaty of Vienna and what WPU's intentions were when they signed it. Also, I hate this name: Worldwide Peoples' Union. This damn organization used to be called World National League or *NU* in the past, I guess... I can't remember the name very well. Anyway, I just wanted to help the good people who live there, especially my mother, that's all," Antony looked a little angry and agitated.

"The Worldwide Peoples' Union is just doing its work, and it only wants what's best for our society."

"Since its creation, this damn group has influenced every government to elect the worst and weakest presidents for their nations; the choice of the people doesn't matter. Also, this damn group just wants the rich countries, with these damn presidents, to splurge on poor countries, like parents who pamper their kids — kids who will become weak and codependent adults. If the WPU is so good, why haven't they talked about the population decline in the last years?"

"What population decline?"

"Don't bullshit me, Mr. President, please," Antony looked a little angry. "In ten years, nobody has been born on this planet, and you know very well about it, and WPU knows very well about it too, and you say nothing."

"I know very well about it, Mr. Sims," Richard looked serious and a little worried. "But I guarantee the best scientists are working on these problems and —"

"The problem is the existence of Cloud."

"No scientist has found proof of any connection between Cloud and the population decline."

"They're all paid by WPU and my family, who would only start worrying about this problem when something freakish and big knocked on their doors. Like what's happening now. You saw with your own eyes."

Richard Sowell took a deep breath before his next question.

"Do you believe in God, Mr. Sims?"

"I hate this question," Antony didn't hesitate to answer.

"You probably don't like to talk about this matter."

"I love to talk about this. I don't like this question. You know —" Antony took a deep breath too. "This question carries a kind of authoritarianism, and its answer can make you the object of a lot of criticism... you know... It's unfair. For me, the right question would be: do you feel free to believe in what you want... and change it when you really want to?

"Hmmm... I realize that freedom is important to you, and so it is for me. In this case, what can we do about Cloud? Shut it down and kill the souls who live there so that people start being born again, even though we don't know if that is the right way to solve the problem?"

"I don't know. To be honest, I don't know," Antony looked doubtful. "I… I admire you, Mr. President, asking questions about God and saying the word *soul*… Since practicing a religion became a crime, people have been afraid to talk about it the way you just did."

"Religions have become a crime only in people's imagination, unfortunately, thanks to the media and unethical nonprofits. I've heard a lot of comments after my accident like: 'hey, where is your God now to cure your legs?'… Well, I lost my legs, but my mind and my heart have gotten stronger every single day," Richard said in a fatherly manner.

Then, the President took a little key from his suit pocket and opened the handcuffs that were keeping Antony at the table.

"What does this mean?" Antony looked surprised.

"Boy… if you don't have the answer to solve this strange situation with Angels, I hope very much you'll find it. However, I'll give you a good tip. In a few minutes, an annoying tall blonde woman will cross this door to disturb you."

Richard pointed to a smaller door on the other side of the room.

"But if you go through that door, you'll find a corridor that will take you to a garage where you'll see a beautiful blue Kawasaki Ninja H5R. I have to admit that I missed my legs when I saw that amazing machine hahaha."

Finally, Antony felt his hands were free, but this feeling wasn't better than the President Sowell's words, which made his eyes tear up.

"Thanks, Mr. President. I'm gonna do my best. I promise."

"I know... and make the most out of the thirty-minute break I gave to the security guards."

***

Outside the wall that surrounded downtown New York, Antony rode his motorcycle freely until he reached the dirty roads, far away from the big city. In the middle of the way, he noticed the Bombardier Global 1K landing softly — he had already programed the aircraft to meet him to expedite this travel. Several homeless people around the place were impressed, watching the jet that seemed to have come from a sci-fi movie. Antony quickly got in the jet through the automatic lateral door, then the jet took off.

Antony looked tired when he got in the jet's cabin, but he was startled by some special visitors.

"WHAT THE HELL ARE YOU DOING HERE?"

One of them replied with a meow. The three moggie cats got in the jet when it was in the old hangar. Antony talked to his friends as he gave some instructions to the jet's AI.

"Well, I don't have time to back to the hangar, but I have some fish-flavored cookies, I guess. AI, set the trajectory for Cloud Base, please."

"Ok, Young Mr. Sims," the AI replied politely.

CHAPTER 26

The afternoon was just beginning in the Nebraska desert as the Bombardier Global 1K flew over it at high speed. The sky was clear, and in the jet, Antony was relaxing in the passengers' cabin, just watching the modern roads and train tracks undergoing maintenance by some giant robots, like the ones he had already seen in the Meno River in Germany but more modern. These fantastic machines always brought back Antony's childhood memories like a deep breath of relief after days filled with challenges. Also, he realized that these machines were controlled by employees inside them and not from a distance like drones. It was a rule when Antony worked in Cloud Base. *There are some problems at work that must be solved in person, in the heat of the moment. Not from a comfortable armchair in an air-conditioned cabin*, he always said in the Sims Corp meetings.

While Antony remembered situations from his childhood and the time when he worked hard in Cloud Base, he caressed one of the cats on his lap. This guy never wanted to have pets because of his hectic lifestyle, but after some years, he was realizing these cats were the only ones that didn't see him as just someone they could take advantage of.

"Hey... I know you will never betray me for fish-flavored cookies. Am I right?" Antony talked to the cats happily.

Another memory that Antony's mind suddenly brought back was about Nivedita. He preferred not to contact her again because Uncle Peter or Christine could do something bad to her if they discovered a call or message between them. There was also Angela; she probably felt very scared when the angels almost destroyed the reactor at Cloud Base. Antony had always avoided meeting his relatives, and he knew that they had been visiting his mother, not because they cared about her, but because they wanted to find him and try to change his mind with stupid orders and moral lessons. Despite hating these relatives' actions, he was sure that the best way to solve Cloud Base's problems was working on the disagreements he had with his family. In short, he had to fix his family before fixing the world.

A few minutes later, Antony caught sight of Cloud Base, a big white heart in the middle of the Nebraska desert, a sight that brought back good and bad memories. Good because Antony learned a lot about technology in this place and with its employees, bad because he discovered that bad people were using all this technology to do bad things.

The news about the return of the Sims Family heir seemed to cause even more fuss in Cloud Base than the fight of the angels in Sector Eleven.

Through the cabin's window, Antony could see some engineers, builders, and robots fixing a big hole made by Abaddon on the roof of Cloud Base. But more impressive than this hole Abaddon had made on the concrete roof was the person standing on the heliport, waiting for Antony — his cousin Christine. The jet retracted its turbines and landed softly on one of the forty heliports.

Antony got out of the jet and faced his cousin, trying to forget the incident in Brussels. Both wanted to make up, but they didn't know how to do it after such a long time hating each other. At least they had a little willpower within themselves to start.

"Well, here we are again," Antony looked downcast and without hope.

"Good to see you again, Antony. I never thought I'd say this," Christine looked downcast too. "So, we can start a tour through Cloud Base for you to know the new installations and... hey... what's that?"

Christine pointed toward the jet, and Antony looked in that direction. Both saw the three cats at the jet's open door like they were employees doing the basic maintenance on the aircraft.

"Humm... Have you ever heard of guard dogs? Well, I have guard cats. Hey! Take very good care of my friends on board, all right?" Antony warned the employees who certainly obeyed him.

Christine let a shy smile escape and saw the joke as an opportunity to start a good conversation with her cousin.

"Come on, Antony, let's talk about business and winged creatures."

"I prefer Angels, and it's better you get used with the term."

The cousins walked together through the corridors, talking about the new technological innovations that had been developed in Cloud Base, business connections between governments and Sims Corp, and some family gossip.

Try to complain less, my son, at least for one day, that's what Angela had said to Antony every Sunday when they met to go to church, and this was what he was trying to do at that moment. Little by little, they felt more comfortable with each other, and for a while, the family strife seemed to vanish.

Another point of their conversation was how important it was to keep Muriel there, far away from reporters, media, and conspiracy theorists. Making a Black Hole appear in the New York city sky seemed to work very well for the supposed Angels' plans. After this last incident, everyone around the world wanted to know about these Angels and the mysterious Black Hole that had appeared and vanished. The theories and discussions about supernatural creatures invading planet Earth took over the internet and especially the imagination of the people — despite the powerful influence Sims Corp had to restrain as much as possible all news about the last strange events.

Another important matter the two cousins talked about was the bigger Angel who had left. However, to figure that out, Antony would have to talk to Muriel.

They approached a large, clean infirmary on the last floor where Muriel was recovering, lying down on a wide bed right in the middle of it. The bed was a little reclined and surrounded by computers, defibrillators, oxygen tanks, and some white tables where pieces of

Muriel's armor and her cracked sword lay. Nobody stayed there except for the special patient, who looked sleepy. Her body was almost completely covered with a big white blanket, and on her face, there were still scars. A large window illuminated the whole place, making the environment seem like a sacred scenery with an injured Angel, her wings broken, trying to heal herself.

"Ok, I will let you two talk," Christine said.

"I know you'll be spying on our conversation through the cameras and microphones hidden in this room."

"You still know the protocols very well."

"No, I know you very well."

Both smiled at each other like old friends, then Christine went back to her office.

Antony walked slowly towards Muriel, who kept her eyes closed for a while. So, to respect her rest, he stood close to a table and started examining a small piece of armor — a piece the size of his hand, silver with blue tones that seemed to shine sometimes. Silence took over this sad place, and Antony's thoughts wandered freely through many doubts: *Where had she come from? Does this kind of metal exist here on Earth? Did she feel pain? Will she continue on her mission from now on?* His fast reasoning was suddenly broken by a sweet and serious voice coming from the bed.

"'Next time we should choose a better place to meet,'" Muriel said calmly. "Only now I understand your comment and what humans call a sense of humor."

"Oh... yes... In New York. I still remember that too," Antony looked calm but sometimes a little nervous close to her. "This kind of metal is...

is very interesting. It probably only exists in your world." Antony held the sample in his hands the whole time.

"Not exactly. Your people can make this kind of metal through a process that you name... metallurgy. It depends on combinations of other elements that your people have in abundance on this planet."

"Wow, that's insane! This seems to be so resistant and strong... sorry... I didn't even ask you how you're feeling. I... I don't have enough words to express my gratitude to you for saving this place. You probably suffered a lot in the battle."

Antony looked worried when he pointed to Muriel's arm that she had almost lost in the battle. Her arm was sewn with thin iron strings, her smashed foot was bandaged, and her wings looked like two branches of a dry tree with only a few feathers. Despite all these wounds, she looked peaceful, without any worries about her body, though Antony could hear how weak she felt the whole time.

"Not so much," Muriel said, "I need some days to heal my body. For me, this is a normal process."

"Wow... and how about the other Angel you fought? Did you know he escaped from here? Maybe he will come back."

"No, he will not come back. We fought, and our God decided that I was the victor; that is it. But Abaddon is an amazing warrior, born to fight; he has a much better and faster healing power than me."

"You... you said God. Do you believe in a God in your world?"

"Like your people in the past. My God lives in another world far away, like yours. But my people are more devoted. My God made the passage of my people to your world possible through a Black Hole; my God is the great responsible for this passage."

"Interesting... so your God can easily open a huge passage and let your people come here anytime, too. Can he?"

"Easily, yes, but not anytime. My God does not want to open a Black Hole... or passage anytime. This is a stipulation my people must follow... we could call it limitations... my people have to face limitations to come here, to complete our mission."

"Your God could help you... and your people more to complete your mission."

"Limitations can help us. Limitations push us to plan our next actions better and give a chance for those who are against us to change their minds. My God knows very well that my people can make mistakes."

"Wow. That's a good way to think. But I still have a doubt about something you said in our first conversation, in France... oh... it's better I let you rest —" Antony looked uncomfortable.

"Do not worry... my Messiah. Feel free to ask me."

"Ok... thanks... In our first conversation, you told me that many different people, including humans, lived together in harmony and suddenly other planets, oops... I mean worlds, appeared, and some people, like yours, decided to live out of here. Correct?"

"Yes, it is true."

"My doubt is... if they lived in harmony, why did they decide to live far away from each other?"

"I hoped you would ask me this question. I will use the same humans' names and concepts for you to understand better... In the past, as I already said, my people and others lived in harmony here on Earth. But something started arising from... from our hearts, or souls... or minds. Something within us."

"Within us?" Antony looked apprehensive.

"Yes, something named... *evil.* A feeling, an illness, a curse. *Evil* little by little was dominating most people. Suddenly, some of them, more powerful and stronger, wanted to dominate the more fragile and weaker ones. One day, other powerful gods, including my God, decided to create other worlds and take different beings to different worlds in order to keep people from enslaving other people and causing a huge war. Then, here on Earth, only the fragile begins were left."

"Humans and animals," Antony completed.

"And humans have stayed here until now, trying to remember how these ancient beings, or gods, were... creating and recreating their visons of them through art, tales, and poems. All of this pushed by a strong will within humans, the will to reconnect with these beings."

"Reconnect? Like religion?"

"Yes, but religion seems to be something your people have lost with time."

"You're right, unfortunately. All of that sounds very sad too."

"I agree with you. Then each kind lived in their own world connected to the *flow of life*, or as your people understand it, a flow of souls and spirits who are born, evolve, and born again in different worlds, a flow which your grandfather interrupted."

"My grandfather only wanted a world where people didn't feel the pain and fear of death."

"And caused more pain, fear, and death. This is what I am trying to warn you and your people. The humans must destroy Cloud. The fight between me and Abaddon was a way to draw your attention to the immense tragedy that can happen. If your people do nothing, a battle more challenging than the last one will happen."

"And can you help us again?" Antony was getting increasingly nervous. "Or can you call some warriors like you to help us?"

"Antony... my Messiah," Muriel looked weaker, "do not try to change my mind... my God let me stay on your side, not to help you and your people... but to drive you to a better way, a better choice. Moreover, I will probably not recover completely for the next battle, when the next and last Black Hole appears."

"And you know when and where this Black Hole will appear?"

"According to the space connection between our worlds, each Black Hole appears in an interval of seven days. It will probably appear close to here."

"Damn... we have about five days."

Antony's heart beat fast, almost jumping out of his chest. A feeling of weakness seemed to take over his whole body; he felt faint but he resisted, and with a trembling mouth he asked:

"Will other warriors like you come?"

"No... my people, by my God's command, will send more powerful warriors than Abaddon... bigger and stronger."

"Bigger and stronger than Abaddon?"

"Yes, like the iron giants that work around this place."

Antony immediately realized that she was talking about the giant robots used to do the maintenance around the Cloud Base. *If Abaddon alone was able to nearly destroy all of my grandad's dreams, imagine what other Angels like him, but bigger, like the robots working around this building, can do.* That's the idea that hit Antony's mind hard.

"Like Abaddon but bigger... Wow... Sounds like a weird coincidence that your warriors would look like the robots here."

"My Messiah... you have a lot of things to learn yet. It is not a coincidence. Your people have more in common with those you named gods than you imagine."

"People who live on Earth and gods? What do you mean?"

"Like I just said, humans still preserve a strong wish to reconnect with gods. They do not have enough power to be like them, but through an extraordinary technological development, they created ways to be at least similar to the beings they lived alongside ages ago."

"I still don't understand."

"It is so simple. Humans already lived in the same world as my people, who they named Angels. Humans would like to fly like Angels, but they cannot do it, so they invented machines like airplanes and helicopters. Humans knew beings who lived under the sea, but they cannot live under the sea. So, they created ships and submarines. They lived with beings who ran at high speeds; ages later, they created speed machines that you named cars, motorcycles, and trains —"

"So, my people lived alongside giants, and after many ages, people created their own giants," Antony was astonished. "So, all creations throughout the ages were made for humans to resemble the gods they had known in the past. That all makes sense now... Wow, the list of gods and extraordinary beings humans lived with is infinite, I guess."

"Yes... here you are gods too. You were considered inferior by other gods in the past, but throughout the ages, your technology has overcome this situation. Your technology has impressed beings from all worlds, except for the one that is destabilizing the universe."

"Cloud... I know. Cloud is proof that humans exceeded their limits. So... as you said, giant warriors will come to destroy this place and. What can my people do to avoid it?"

"Your people can do everything in their power, but the destruction of this place is inevitable. If your people stop it the first time, then there will be more battles, more difficult ones, and more, and more until this place vanishes."

"VANISH, no!" Antony's eyes showed fear. "But you saved this place once and —"

Despite her weakness, Muriel started expressing more seriousness, like a wise professor who has lost her patience with her student.

"I did not save this place. Indeed, I gave you and humans in general more time to think and understand all the destruction that this place is causing. If your people do nothing or keep stopping my people from destroying Cloud... my people, following orders from my God, will be obligated to destroy not only this place but your whole world as well."

"MY WHOLE WORLD. What do you mean? Your mission wasn't just to destroy this place? Now you're threatening my planet," a feeling of despair took over Antony's face and he was sweating nonstop.

"My Messiah... do not try to change my mind. I just explain the truth. If your people destroy Cloud, the equilibrium of the universe will be saved, but if my people destroy it and let your people free... your people will build another Cloud, and another one, and another one... and my mission will have been a big failure. Then, if my people destroy Cloud, they will be obligated to destroy the reason for its existence."

"Human beings." Tears mixed with the sweat on Antony's face.

"Yes... the true mission given to us by my God was not to destroy the humans' terrible invention, but to destroy the human mentality that led to its creation."

Antony felt his heart beating fast and his mind ready to explode, like a furious volcano. An animal instinct seemed to take over his subsequent actions, causing him to just drop the piece of metal on the nearest table and run out of the infirmary — tormented and lost.

Muriel just watched her host leave quickly, like a scared prey escaping from a savage hunter. Her face, that looked peaceful, in that moment expressed some melancholy, as someone who had just uttered what must be said, no matter how deeply it could hurt.

***

The sound of someone throwing up echoed through a long corridor in Cloud Base as Christine approached the wide door of a restroom. The noise was still intense when the general manager carefully got in the well-lit place and noticed her cousin with his head in the last toilet in the back.

The restroom was very clean, spacious, and white, with some aluminum details and large mirrors.

Christine slowly got closer to her cousin, who didn't notice her presence for a while. She looked worried, and with a whisper, she tried to draw his attention.

"Antony."

Her cousin got startled and moved abruptly away from toilet. He stayed sitting down on the ground between the door and a sink, and his face showed how desperate he was.

"Christine... you heard... you heard what Muriel said —"

"I heard everything, Antony. Come on, let me help you," Christine reached out.

"Yes, yes, you heard, I know. I'm so stupid. But just leave me here... I need to think of some solution to all of that," Antony refused his cousin's hand.

"Ok, but... there are other places you can use to think about how to save the planet better than a restroom smelling of vomit. And you could have used the sink."

"You're always trying to help me," Antony sounded sarcastic.

"Ah... Sorry, Antony, but... I mean, you're not alone," Christine looked a little worried remined kind. "We can solve this together. We have already solved many problems together. Look at the resources that we have."

"You're not afraid?"

"Yes... yes, I am... but we're safe here."

"I'm not talking about us but about our planet. Nobody was born in the last years, and probably in the near future, there will be few people while most of the population lives happily in this big damn USB drive. Moreover, now evil Angels want to destroy the remaining humans who will be here! Can you see? We're in the middle of a silent and slow extinction process... that can accelerate in a few days."

"Your prediction seems from the WPU's reports. You remember? Melting of polar ice caps, the ozone hole that is going large, the end of the oil age. You always thought these theories were stupid, and... look! You were right all the time. They never happened."

"I know, I know, but now it's different. You can see with your own eyes in the infirmary. You can go in the infirmary —"

"Antony, take a rest at least, it's already night and —"

"No, we don't have time. I must go back to the infirmary and convince her to help us save Cloud... to save the planet and —"

"Antony, did you hear what the monster in the infirmary said? She threatened us, she threatened our planet. Do you think you can ask for the help of someone who wants to kill you? Kill us?"

"She saved this whole place —"

"One time... what will she do from now on? A friend of hers sent hundreds of armed men to Cloud in only a few minutes, leaving an immense blood trail, and she talks about him like he was her boyfriend. She's a menace, Antony, you heard it from her own mouth."

"I... I have to think more about all of this, leave me alone... please," Antony didn't have the courage to look directly at Christine.

"Ok. But, please... don't try anything foolish."

Christine slowly walked toward the door, but probably following a sudden instinct, she stopped and spoke like she was in a confessional.

"Antony... remember Grandad Allan just wanted a world where people didn't have to feel the pain of death anymore... and he got it, and het got a beautiful place for Aunt Angela, too."

Antony kept silent.

"But there is a... a winged monster downstairs who can destroy all of that. Our grandad's dream, your mom's home... We can't let it happen," then Christine left the restroom.

After a few minutes of silence, trembling like a fragile animal hiding from a hunter, Antony left the restroom and entered a corridor accessible only to right authorities of Cloud Base could access. Then he went into his private bedroom where, after a long bath and taking some sleeping pills, he fell asleep. Before his eyes closed completely, he remembered the nights in Brussels when he fell asleep every single night, just wishing to forget the existence of the world, at least for some hours.

CHAPTER 27

In Christine's office, both cousins were looking down, completely lost, thinking about what Muriel had said only a few minutes ago. Christine sat in her armchair, while Antony sat in another one in front of her. Between them was a wide desk, its top like a large touch screen which showed several images of Cloud Base's sectors and Muriel lying down in the infirmary. Two wide screens above the cousins were ready for the beginning of a meeting with two important political leaders.

"Well… you heard what Muriel said. Any idea how to protect this place?" Antony looked worried as he asked.

"For now, I've given orders to some more soldiers, about five thousand humans and small robots. They will get here tomorrow."

"More soldiers? You saw the security cameras' footage of the battle. Just one Angel killed more than fifty men with two boomerangs. Imagine what a hundred feet tall Angel can do."

"I told you that there was time enough to transfer all these soldiers' minds to good places in Cloud the moment their bodies were killed. This is part of their life insurance," Christine spoke calmly, but she looked worried as well.

"Yes, you told me. But that's not the question. The question is Abaddon was just a *free sample* of what these Angels can do. Now we are sure that someone worse than him is coming."

"Maybe she was just trying to scare us. But anyway, I've ordered some low-intensity atomic missiles from our factory in Texas too.

Without authorization from the government, of course. They would forbid it or bureaucratize our actions, probably."

"No, don't you realize an atomic attack can affect Cloud Base? If the radiation gets in, it can affect the whole memory system."

"Ok, but I'm just doing what is within my reach. If you don't give me some ideas too, it will be harder to solve this problem."

"Yes, I know... but... oh... nothing comes to my mind right now."

This was probably the first time Antony felt affection for his cousin and agreed with her. The empty feeling that hit him after Muriel talked about the destruction of the planet still dominated his mind, which was used to solving any problem in a few minutes. At least this new beginning with Christine was bringing him good emotions. *Try to complain less, my son, at least for one day*. His mother's advice was working in practice.

Antony decided to forget his last conflict with Christine and Uncle Peter in Brussels. *They just wanted to keep me out of trouble*, he thought. An egoistic attitude from his relatives, but maybe induced by Yelena. This tug of war between corporations and politicians to gain power was brutal and unfair most of the time, and Antony started realizing that Christine's insistence in this game should be valued and his escape could be seen as a show of weakness.

On Christine's desk there were some Headgates and a lounge chair, like the one Antony had in his apartment in Brussels. He took one of the Headgates and asked:

"Can I keep this one with me? I didn't find any of them in the jet. Also, I have enough time to visit my mother before the meeting, I guess."

"Of course, no problem. We have one hour until the meeting. But you know you might meet Uncle Peter there, don't you?"

"Yes, I do, but for me, who has faced Angels that want to destroy the planet, facing Uncle Peter will be easy."

"Humm, good boy. Ok, I'm gonna check out sector eleven. I didn't think blood would be so hard to clean. See you soon."

"See you soon."

Antony lay down on the lounge chair with the Headgate on his forehead as Christine left the office.

"Ok, here we go again!"

CHAPTER 28

Inside the red brick church, just known as *the church* since it was the only one in Cloud, Angela was kneeling, praying in her favorite spot — in front of the strange stone sculpture of the book on a stand and looking at the sculpture of the *Virgin and Child,* by Giovanni Pisano. Only Angela had attended this church since she arrived there. This situation made her a little sad and solitary because going to this place reminded her of people's lack of faith, who had forgotten their religions thanks to a perfect world full of pleasure.

Walking on the street toward her home, Angela thought about this feeling of sadness and solitude that suddenly hit her. *People believe in God only when they are suffering, but when they are full of futile pleasures, they forget their religion completely.* This was the idea that floated around in her mind and made her sad as children played happily along the street. However, an unexpected vision changed her mood and brought a smile to her face. Antony, who looked a little down as he beckoned to her, stood in front of her home. Since the incident of the lights in the sky that scared everyone in Cloud, she had been worried about her son.

"Antony, are you ok?"

"Hi, Mom. Yes, I am... of course. I'm better now that you're here."

He hugged his mother in a gawky way, smiling shyly.

"It's amazing. I couldn't bear to wait for you until Sunday."

"Yes, me too."

"Yes, me too AHAHAHA," a joyful voice rang loudly behind them.

Mother and son looked a little scared as they turned around at the same time and saw Uncle Peter, who appeared on Angela's porch. He laughed a lot, holding a pack of beers as he walked toward his two relatives, hugging them in a friendly manner.

"Hey, how are my favorite relatives doing?"

"Oh, this day has been amazing with you two showing up," Angela looked blissful.

"Yes, if Uncle Peter lets us breathe," Antony looked gawky.

Uncle Peter was a corpulent man, taller than the two. He seemed always to be attending a party wherever he went, especially in quiet places. His sudden appearance made Angela very happy because she couldn't remember the last time she met Antony with other members of the family. She led the two visitors affectionately into her home.

Antony still looked gawky as Angela and Uncle Peter talked happily about funny situations from the past, travels, and some unforgettable family parties — especially the situations that included Antony.

Everyone gathered in the ample kitchen of Angela's home. This part of the home was illuminated by wide windows and a skylight. Its walls had light green, pink, and white tones — all decorated with lovely pictures of landscapes, clocks, and hangers where all the kitchen utensils were suspended. It was a typical grandma kitchen that looked like a place created for a fairy tale, with a round wooden table where the three members of the Sims Family were enjoying a great snack with coffee, cookies, juice, cakes, and the beers brought by Uncle Peter. Angela scolded her brother for having bringing brought beer, but after a few minutes of conversation, she was convinced that nobody would get a hangover there, unless the Cloud inhabitant asked some programmers outside to allow it.

Uncle Peter could cheer people up in any situation, always emotional and gesturing a lot. Each family story he remembered was told through a funny narrative or sarcasm, like when Antony was eight years old and invaded a bank's computer to take money to buy ice cream. The cops surrounded the mansion to capture the *dangerous hacker*. Or when Peter took all teenagers of the family, about nine drunk, crazy guys, including Antony and Christine, to a midnight party on the beach in the Orange County, California. Everyone decided to race each other naked.

"Hey, Angela, it was hard to pick the whitest ass in that race ahahah," Peter spoke and laughed at the same time.

"Yes, I lost my shorts in this race, and I came back home trying to cover myself up with one of Christine's T-shirts. My shorts are probably still lost on that beach now ahahah," Antony spoke and laughed too.

For a moment, Antony forgot that an hour ago an Angel from another world had threatened to destroy the planet, and Uncle Peter wanted to keep him on house arrest in his apartment in Brussels. He still remembered the way his uncle had treated him in that situation, but Antony didn't want to talk about that in front of his mother. *Maybe my uncle was pretending to support Yelena, the main villain of this story, and protected me,* he wondered. The feeling that arose in this family moment made Antony give up solving the family disaffections and let good vibes take over this meeting, especially when he realized how happy his mother was. There wasn't a mother in the world who didn't want to see all members of her family happy together, and Antony started feeling he was responsible for his mother's solitude in her routine between church and home.

Before getting to Angela's home, Uncle Peter made a quick tour in other places of Cloud, and it was the other subject he pushed into the conversation.

"Look at this incredible place... in one hour I went to the coast of Monaco, parks of New Zealand, and beaches of Brazil. And all of it just lying down in a bed at my home in California. Well, Angela, our father gave humanity an amazing present. I haven't seen poverty, violence, or sad people here. I have never felt so free, and I confess I'm excited to live here forever. If you understand me," Uncle Peter spoke happily, holding a beer.

"Aren't you exaggerating, Uncle?" Antony inquired, doubtful. "Well, people here don't make any effort for anything, and they have what they want so easily. I think if people don't make an effort to get what they want, they don't value it."

"Who are you to speak about effort, Antony?" Angela intervened seriously. "You were born gifted into the richest family in the world. You're brilliant in anything you do, without needing to make much of an effort."

"Yes, Angela. Show Tony who is the boss. Hahaha... And look, boy. It's not an exaggeration. Imagine a person who had an accident and lost their movements or someone who was born to a poor family and didn't reach their goals. Anyway, imagine a person who made a lot of mistakes and now is too old to begin again. But here... here it's different. Here, they can have the life they always wanted. Here these people have a second chance. A chance that God didn't give them, but your grandfather did."

"For me, God allowed my father to build this place," Angela said. "So, God gave humanity a second chance to live the way they had always wanted out of here."

"Ok, ok, you can interpret it this way. Summing it all up, if someone wasn't born rich, healthy, or gifted they can enjoy a good life here after a hard life outside."

Antony had no other option except to hear and accept the justifications of his mother and uncle. He had never seen Cloud like as a second option for people who hadn't had an opportunity to build a good life in reality, and in that moment, he realized that he was more than privileged — someone who had required effort from others. Also, he remembered President Sowell and the problems he had to endure even to move in a wheelchair. This man, more than anyone else, deserved to have a good life after his term.

Most of the time, in our minds, we demand more from our friends, parents, and partners, but we forget to demand more from ourselves, as if we are perfect people who never need to change. These were the ideas that were floating through Antony's mind after he paid attention to Angela and Peter's conversation.

"Well, I have to go," Antony said, "I have an important meeting with Christine shortly," Antony looked a little down.

"It's very nice to know you and Christine are working together," Angela said happily.

"Yes, these guys have to save the world," Uncle Peter completed, "I'll follow you to the door."

"Uncle, we're in Cloud. We can go back outside from here."

"Come on, Tony, just pretend to live a normal life here for once."

"Ok... sorry, Uncle."

Uncle and nephew said goodbye to Angela, who looked tired.

Despite people not having physical bodies to feel tired, their minds needed to relax sometimes — just like people outside need to take a rest. While Angela went to sleep in her bedroom, Antony and Peter went out slowly, until they stood in front of the house.

"The weather is fantastic today," Uncle Peter was excitedly looking up at the sky, "I'm gonna go to the Swiss Alps, then I'll play volleyball in the Gold Coast, in Australia. Would you like to come with me?"

"I would like it very much, but I have a meeting with Christine and other people to talk about... you know... saving the world."

"Oh, sorry... I just forgot. But don't worry, everything will be well soon enough. Like your mother said: If God allowed your grandfather to build all this magnific place, why would He send monsters to destroy it? Come on."

"Yes, you could be right... maybe."

"I'm always right... hahaha... Look, next week I'm gonna organize some parties in my mansion in, California. Just real life, not in Cloud. I know you prefer real life and real girls too, hahaha," Uncle Peter patted Antony's shoulder in a friendly manner.

"If I'm alive and the world too... maybe."

"Stop saying *maybe* all the time. You know it's a fear symptom, and very dangerous too."

"Fear symptom? What do you mean?"

"Listen to me, and please don't get me wrong. You've hated your own family for a long time, and you got used to this idea. Now you're incapable of being reconciled to us because of your ego, because your mind understands that it's normal to hate your own relatives. To be honest... you can't imagine yourself living happily with your own family. Just try it, and it'll probably feel strange even to yourself."

His uncle's words went straight to Antony's heart and mind, a mix of frustration and shame hard to express immediately.

"Well, maybe... I mean, yes. I should think more about it. You're right... absolutely right."

"My boy, you're improving every single day. Now, I have to go. Feel free to drop by in California."

Uncle Perter hugged his nephew with one arm while he patted Antony's chest. Then he said, *time to go*, probably the sentence that he had programmed his system to understand as a command to take him out of Cloud, because he vanished a second later.

Antony kept sitting in front of his mother's home, thinking about the hard lesson he had just heard. He completely agreed with Peter when he said *you got used to hating your family*. He tried to imagine himself happy with his relatives, but it was such a strange thought, which confirmed his uncle's theory. Despite Peter having bought almost all of his degrees, his power of persuasion and confidence were great, and no university could teach it.

On the streets and in front of the homes, several kids enjoyed good moments playing baseball, basketball, and riding their bikes with their parents. Antony looked very sad when noticed it, then he vanished.

CHAPTER 29

In Christine Sims' office, both cousins were waiting for two important politicians to appear on the screens above them to start a meeting. Christine sat in her armchair and Antony sat in front of her.

Suddenly, there was a discreet signal ring from the left screen, and it showed the video call of Yelena Orlyk, looking imposing.

"Good afternoon, Sims Family. We have to start this meeting right now. It's an emergency that requires every second of our efforts."

"I agree, but the presence of the President of the United States is as important as yours," Christine said.

"Despite Cloud Base being located in America, most of the shareholders and investors are from Europe. In my opinion, Mr. Sowell should be more careful with our nation instead of being late for an important meeting and letting some delinquents run away with no explanations," Yelena looked at Antony.

"I know the actions of my cousin were a little troublesome," Christine spoke pointedly, "but if he hadn't acted that way, the main subject of our meeting today would be 'how to rebuild Cloud.'"

"Also, he didn't run away, Prime Minister; he was set free by an important authority of America." Suddenly, President Sowell appeared on the screen to the right.

"So, Mr. President," Yelena didn't look a bit embarrassed, "you were secretly listening to comments from other authorities. According to the law, that can be considered espionage, which is also a crime."

"Other crimes have been committed by you in my country, but I chose to forget it," Richard replied seriously.

"Sorry, gentlemen... ladies... guys," Antony looked shy. "We only have four days to solve a problem that we still don't know the size of. It's better we keep our focus."

"I agree," Christine said, taking a deep breath.

For three hours, they talked about a lot of strategies to protect Cloud Base. What kind of weapons could be used? How many soldiers, drones, and robots America and Europe could collaborate with in this battle? And other matters too. Richard Sowell ordered all American soldiers to be ready to attack, but he would send only five thousand soldiers to be organized in camps around Cloud Base, and Yelena was ready to send thirty thousand jets and seven thousand tanks, but it was hard to confirm that all of them would get to America in four days.

Another problem they discussed in the meeting was how the footage of Abaddon killing so many soldiers and the fight between the two Angels was somehow mysteriously leaked on the internet and made a lot of noise in society. Antony swore that he had nothing to do with this because he was in jail when the battle happened. *Any employee could have done it without anybody noticing,* Christine affirmed. It could have happened in any other situation but this one was causing fear among soldiers and their families. Despite *death* being something that had been overcome, the possibility of Cloud Base being destroyed was shaking the faith of humanity. The mystery of the images being fake or not gave rise to more and more conspiracy theories, which gained strength in people's imagination.

For President Sowell, the priority was to preserve soldiers' lives in reality. He could organize a huge army to fight against the Angels and eliminate them quickly, but if the beings coming next were as strong as Abaddon, a carnage could happen — and the American leader could be the one responsible for it. For Prime Minister Yelena, the priority was to protect Cloud Base and her unethical business with Sims Corp, but sending an army from a continent that had a low number of inhabitants could be a risk. She had faced many problems with immigrants and battles on the frontiers of Europe — letting too many soldiers and weapons get to an indefinite war in America would make her continent vulnerable.

"Ok, this is what we have for the moment," Richard led the conversation, "ten thousand soldiers and more on the way from Europe. I'll allow the use of atomic missiles in combat."

"But only to be used as a last resort," Yelena completed.

"However, tanks, cannons, and heavy artillery will be totally allowed."

Antony and Christine nodded to each other in silence in a gesture of confidence.

"Now, we only have to wait for a challenger that we'll probably overcome," Yelena said, "of course with the best help from our partners at Sims Corp, especially the genius of Antony. Well, I need to manage a situation here on my continent, then, as soon as possible, the European soldiers will be in American territory. Goodbye."

"Keep faith, guys. I'll be there as soon as possible as well. We'll keep in touch."

Both screens automatically turned off their transmissions. Christine looked more relaxed, and Antony was down and quiet — an uncommon behavior from him.

"What's the matter, Antony? You should be happy about this meeting. They agreed with every suggestion we gave."

"I don't know, for me, all soldiers in the world completely armed can't protect this place."

"So, why didn't you talk about it with the President and Prime Minister?"

"And say what? That I have a bad felling?"

"Your bad feelings have helped up to now."

"But now it's different, it's reality. I need to present powerful arguments to them. We're risking people's lives; it doesn't matter if they will go to Cloud. We can't treat life as something disposable that we just throw into Cloud."

"There'll be a lot of drones and robots working together with humans. The soldiers will be fighting safely from their cabins around Cloud Base. Look, Antony… now we are working together, and we have at least some notion of what is coming."

Antony took a deep breath and kept looking up doubtfully as Christine continued speaking.

"You look different. You'd usually argue sarcastically with political leaders and corporations' presidents. Now, you just agree and seem to retract your typical explosive behavior."

"To be honest, I should get some sleep. This day has been full of emotions. Thanks for defending me from Yelena in the beginning of the meeting. I thought you always wanted to be like her."

"No way," Christine looked disgusted, "she acts as if she were the last woman in the world. Did you notice her hairstyle? I still remember you said once that her hair looked like a nest built by drunk birds. I can't be serious when I remember that."

"Wow, remind me to give you a joke book for your next birthday. I haven't seen you laughing this hard in a long time."

Their moods both changed completely, and Antony started feeling differently about his family since he'd left his mother's home.

They stayed there for a few minutes just remembering old jokes and laughing a lot, like teenagers on a school break. Then Antony went to his bedroom, and Christine stayed in her office working.

***

It was 5:00 a.m., and Antony was lying down on his bed, unable to sleep. The bedroom was ample and modern without many decorations. Just a king-size bed, an armchair, and a wide screen. A large window, curtains wide open, occupied a whole wall, providing a beautiful view of the Nebraska desert. Despite the room having white and beige tones, it was a gloomy place, only the weak light from a few stars and a faraway orange horizon gave it some color. After a day full of emotions, threats, and reconciliations, Antony couldn't keep his eyes closed any longer. *Life seemed easier when I just ignored all the world's problems, like most people,*" he thought. In that moment, his responsibility towards the world became bigger than he could bear alone.

Antony got up, approached the window, leaned his hands against the glass, and stayed there, just watching the orange line on the horizon ahead. He was wearing pajama pants and a tank top that showed clearly the big scar that crossed his chest. His face showed doubt and sadness, a mix of feelings that, in these last days, insisted on asking for an answer, but he still didn't know what he really wanted.

Giant robots stood next to the roads and train tracks, and several buses, cars and trains went to and from Cloud Base. Antony realized that this place couldn't stop because there were billions of people who depended on it. Not just out there, but especially inside, including Angela. The robots were being softly illuminated by the first lights of sunrise, bringing back to Antony's mind a crazy idea he'd had a few days before, but that, in these decisive next few days, he could put into practice. He quickly put on his nylon coat, jeans, and tennis shoes, then he got out of the bedroom and went toward the infirmary.

The corridor leading to the infirmary was dark, empty, and quiet. Antony walked carefully into the room where Muriel was still sleeping, her body covered with a white blanket. He quietly approached the table where a few pieces of Muriel's armor lay exposed, then he took a piece of metal that he had observed the day before. Antony put it in his coat pocket and went silently toward the exit, but before he could leave, Muriel's weak voice made him freeze for a moment.

"My Messiah... you know, it is inevitable."

Antony slowly turned to her and said in a sad voice:

"They're my family... and this is my world. I can't accept that there is nothing else to do."

"But you must accept it. Imagine a wolf mother and her cub. She teaches her cub not to fight against animals that are bigger than him, but he disobeys, and he fights, and he dies... the wolf mother is sad for a long time, and later on, she gives birth to another cub... She teaches the same lesson to her new cub, and this one obeys, he grows up and becomes a strong wolf... this way, the circle of life is restored, but it cost an unnecessary sadness and time... To keep the circle of life, it is not necessary to control it but to learn about it... So, my Messiah, the destruction of Cloud is inevitable, but you can avoid the death of millions of your people here and beyond."

"Beyond? I don't understand you."

"Destroying Cloud and avoiding the death of millions of people here and throughout the universe."

"Avoiding the death of millions of people by killing the people who are living inside Cloud. I don't think so," Antony started seeing Muriel as an enemy.

"I have known you since you were a child... you have always disagreed with the path your world is taking... So, if you had the power to fix it... would you do it?"

"And if your people lose the next battle against my people, would you leave? The way Abaddon did?"

After a moment of silence, probably caused by her feeling weak and doubtful, she finally said:

"There will be destruction. Your Cloud or your world. This is inevitable."

"You didn't answer me."

"You did not listen to me anymore, my Messiah."

"I'm not your messiah."

Antony looked outraged when he turned and left the room without an exact answer. An empty sensation took over his mind and he was hating it.

Then he kept on his way and left the infirmary, leaving the patient alone.

This was the first time that Muriel's face expressed a different feeling. And this feeling was sadness.

***

In another bedroom, Christine got up to get a glass of water from a jug on the nightstand beside her. She was wearing a black silk pajamas, looking sleepless. Her bedroom was bigger and more luxurious than the one where Antony had been sleeping, with many shelves where there were several books and a small desk with a computer on it. A wide window occupied a whole wall there too, but its curtains were almost completely closed.

Christine noticed a different craft flying away from Cloud Base through a gap in the curtains — a dark blue Bombardier Global 1K could be seen in the horizon that was gradually becoming orange. She watched the aircraft vanishing in the horizon seriously, not surprised. *It's working... just like I planned,* she whispered.

CHAPTER 30

Border between Kansas and Nebraska.

In a huge green area located between Belleville in Kansas and a village called Chester in Nebraska, there was poor but organized community formed by cracked wooden houses, some schools, churches, markets, and large, tall buildings that looked like a combination of a barn and dockyard made out of wood and metal. The locals often referred to these places as *Dockbots*. All these places were separated by simple gardens, small sand fields where kids and teenagers played soccer every single day, and sandy streets — like a place that combined the features of an abandoned little town and a medieval village, with some large pieces of scrap metal that were just used by kids as playground toys. The variety of ethnicities was remarkable in this place that was just called a community. Most of the people were black Americans, Latin Americans, Asians, some Jewish, and Irish. Both adults and teenagers were often wearing dirty mechanic's uniforms.

Out of one of these *Dockbots* was coming a lot of noise and smoke. The area was surrounded by bushes, tree trunks used as park benches, big abandoned tires, and other rusty pieces of vehicles. There were some curious children sitting down on tree trunks and tires, just watching the *Dockbot* shake like a blender, and waiting to see what could get out of there. This place has about eighty feet tall and thirty feet wide, with some big pipes and gears on its walls. Its massive wooden gates looked like those of a medieval castle, and they were shaking a lot because something inside was emitting a motor roar, as interesting as it was terrifying. On the gates, two big letters were painted: 'G J'. This *Dockbot* seemed more and more alive every second, ready to move at any moment, and the children around it were calm, as if this was a common occurrence in this community.

Suddenly, the gates opened, and out emerged a large robot, several shades of gray and towering sixty feet tall. It was easy to identify many different pieces of vehicles — trucks, buses, cranes, and ships — that together formed this imposing machine that looked like a metal giant. At the center of the chest and head area, there was a kind of tractor cabin where a confident woman stood, piloting this giant. She was standing up in the cabin as her arms controlled the robot's arms, while her feet controlled the robot's legs through a kind of pedal below. Only two thick, chest-high horizontal metal bars and two other vertical bars, one on each side, protected the pilot. Some thinner metal bars protected the pilot's legs.

When the robot left the *Dockbo,t* all children around cheered as if they were in a Super Bowl final or in a World Cup. They all followed the metal giant, shouting out loud repeatedly one name: *Glenda, Glenda, Glenda*.

Glenda Jackson was a tall, strong black woman who was wearing worn-out denim overalls and a dirty yellow t-shirt. Her head was shaved, which highlighted her silver earrings and thick lips that always had an intense red lipstick on — a pretty face like a doll but with a brave expression like a soldier in battle. On her wrist, Glenda always had silver bracelets with a special detail. Despite being in her forties, Glenda had an athletic body and wide shoulders, which enhanced an aspect as imposing as the robot she was piloting.

Glenda piloted her machine down the streets of the community, followed by kids riding bikes, running, and piloting little robots like hers, but which were five or six feet tall. Everybody was excited, shouting her name. The people on their porches and windows cheered her, yelling things like *finish him* and *kick his damn fat ass*. This was the typical ritual for when a fighter was going to enter a fight ring, and Glenda was always the favorite.

After a couple minutes' walk, Glenda and the kids arrived at an open area in the outskirts. A huge hole that looked more like a crater, surrounded by wooden bleachers protected by tall fence made out of metal cables, formed a wide sand arena. Next to this place there, was a big *Dockbot* where seven robots about a hundred feet tall were just standing as if parked there.

Glenda Jackson got in the arena through one of the accesses, beside which there was a traffic board displayed 'The Maximus Arena' written on it. She arrived, bravely raising her robot arms bravely in a gesture of thanks to her fans. This area wasn't full, but the small crowd that was there screamed and cheered a lot. From the access on the opposite side, another robot, similar to Glenda's, got in moving its arms like a kung-fu fighter, trying to intimidate its opponent. Its pilot was Mike Hong, a Chinese man in his fifties, but brawny and an expert in machines and martial arts. His gray hair and scars on his arms and face were proof that he had a long experience in fights. However, all the support from the crowd was directed at Glenda.

"Come on, nigga, it's best you go back home take care of your kid than be embarrassed in front of everybody here," Mike said out loud in a sarcastic voice.

"I'm gonna go back home after making you swallow your own teeth through your fat ass," Glenda replied in a defiant voice.

The exchange of insults made the crowd even more vibrant, eagerly anticipating the best fight ever.

Both robots positioned themselves to attack, as the people around them shouted out *fight,* in a unique and wild chorus.

On one of the bleachers, there was a big rusty gong held by thick chains attached to an iron arch. A man wearing a mechanic's uniform hit it with a long hammer, signaling for the fight to begin. The sound of the gong echoed amid the screams of the crowd.

In the beginning, the two fighters walked sideways, just carefully studying carefully the best way to hit their opponent. Each robot's steps raised dust, highlighting the first movements inside the arena, like a dance of big barbarian warriors. Mike advanced quickly, executing rapid kung-fu moves known as Lat Sao, but Glenda skillfully defended herself. She adeptly dodged Mike's punches three times until she got an opening and hit his knee hard, causing him to lose his balance. Glenda took the opportunity to seize one of Mike's arms and throw him to the ground. Then, she tried to punch him, but he quickly got up fast — causing Glenda to hit only the ground.

The crowd went wild with each strike and movement of the robots as the fighters breathed rapidly, not even blinking. They were both good, skillful, and strong, so the fight would be determined by details and the best strategy.

"Give up, Glenda; spare your son the sight of seeing you defeated on the ground."

Glenda saw her son, William, on the bleachers. He was an eighteen-year-old boy, very excited to see his mother fighting. Also, he never stopped cheering her on, yelling the entire time sentences like: *Finish this jackass, Mom!* or *Give him the air kick!* — the most difficult strike that only Glenda was an expert in.

"Now you've made me notice my son is here... I'm gonna finish you with even more pleasure."

"So, show me, crazy cow!"

Glenda advanced towards Mike, pretending to punch him, but when he defended himself, she dodged and spun quickly at the same time. It gave her a chance to kick him in that same knee again. Mike's leg lost several screws, and some hydraulic hoses were broken, squirting oil and letting out smoke. But it wasn't enough to knock Mike down, who punched Glenda back several times. One of Mike's punches hit Glenda's robot's waist axis hard, making it lose its stability.

Glenda fell down on the fence that separated the bleachers and the fighting area. Mike hit another punch on the bar that protected Glenda inside the robot, but she got away quickly before her opponent started another sequence of punches.

"You're fast, but it isn't enough… crazy Chinese. You hit like a baby, bro. Your punches wouldn't even be enough to make bread."

"You talk a lot for an animal. So come and show me your best punch… if you can."

Glenda knew that Mike's leg was compromised, and he couldn't move quickly. If she hit him, he wouldn't be able to get back up. Because of that, he insisted on pushing Glenda to attack him. The waist and arms of Glenda's robot were compromised too. So, her next attack would probably be the last one, and it had to be perfect.

She approached her opponent in short zigzag movements as Mike was ready to punch. When he felt the opportunity, he punched toward Glenda, but she dodged and threw the body of her robot to side with a spin. Someone from the bleachers screamed, *It's air kick time!* The movement that preceded this difficult strike consisted of a jump and spinning the body of the robot.

For a short moment, Glenda's robot seemed to float in the air as her heel hit hard the bar that protected Mike. The shock was so intense that it twisted the bar and made Mike's robot lose its balance and fall. The robot hit the ground hard, making people go even crazier.

After the ultimate kick, Glenda's robot stood, and she just watched her opponent lie on the ground, trying to get up but unsuccessfully — his robot's broken knee made it impossible.

The success of the *air kick* depended on how the pilot distributed the robot's weight in the moment of the jump and the position of the opponent, and Glenda knew how to do it with mastery. She was breathing deeply when she approached Mike and stood close to him. The crowd didn't stop screaming from the bleachers, *Go Glenda!* and *No mercy!* waiting for the last strike that would finish this great show. Mike was sure that the fight was lost and said in a puffy voice:

"Come on, woman, give me your coup de grâce."

"No, bro. You were fantastic," Glenda spoke friendly. "The *air kick* was my last chance. If I had failed, I wouldn't have been able to carry on... and certainly, you would have won."

Glenda reached out her robot hand to Mike, who took it firmly, then she pulled and helped him up. The crowd clapped, screamed, and whistled a lot, making a noise that could be heard throughout the whole community.

"Sorry for involving your son in my threats," Mike said.

"No problem, bro. This is part of the game," Glenda winked.

The big robots rose their hands together, thanking the noisy crowd on the bleachers. Glenda and Mike were smiling when suddenly something up in the sky changed Glenda's mood. She looked suspicious and angry at the same time.

"What the hell is that?" Glenda whispered.

"Do you know what they want here?" Mike asked doubtfully.

"I know just one thing, bro... they only bring problems."

Both robots kept their hands up as their pilots just watched a dark blue Bombardier Global 1K, with a white infinity symbol on its fin, landing near the arena.

CHAPTER 31

After landing his jet close to a *Dockbot* and a few robots, Antony Sims walked slowly, with his hands in his coat pockets, on a sand field toward The Maximus Arena. Glenda Jackson, leading some guys from the community, walked slowly toward him. In the middle of the field, both stopped and stood, suspiciously watching each other. A few tense seconds later, which seemed like an eternity, Antony started talking.

"I could say 'take me to your leader,' but you're already here," Antony looked likable. "Glenda Jackson, you never lose your imposing ways. How are you?"

"I was better before I saw your jet landing here. So, get your fat ass out of here."

"I need your help, and I can pay for it."

"You can pay for everything but my help. So, please, get out of here."

"You don't know what kind of help I need yet."

"Doesn't matter, I can't help you," Glenda turned away from him.

"Even knowing that the whole planet is in danger of being destroyed in a few days?"

Glenda stopped and turned to Antony again. The men behind her looked suspicious and stayed on the field, just watching the strange visitor.

"Well, if you live here with your family and friends," Antony continued, "let me explain better. If you live on this planet, and if this planet is destroyed, you and your family and friends will be destroyed too."

"What do you mean? I've already done business with you, and I don't want to anymore."

"You've never done business with me."

"I did with your damn family. It's the same thing."

"No, it isn't. You're famous for never running away from a challenge. At least let me tell you about the situation."

Glenda looked down for a moment and took a deep breath. Then she looked at the men behind her to dismiss them.

"Alright, bros. No problem. Let me take care of this playboy."

Then she looked at Antony.

"Look, I don't have champagne or caviar like you're used to in your mansion. I hope you're not a fussy boy."

"Caviar for me is trash. Black coffee is ok, if you have some, of course."

"Are those your flight attendants?" Glenda pointed to the blue jet in the background.

Antony turned back a little and saw the three moggie cats leaving his jet through the open door. Some kids approached them and started playing with the newcomers.

"Yes... they are the best employees," Antony said, smiling.

"OK, follow me," Glenda nodded and walked on, keeping a serious expression.

***

Some teenagers and kids played soccer in a sand field in front of Glenda's home, a cracked wooden house that looked abandoned, like all other houses here. There were about ten guys having fun on the field, including Glenda's son, William, and there wasn't a goalkeeper because the goals were too small, measuring 3 ft x 3 ft, made out of rusty pipes. The kids were full of energy and played like there was no tomorrow, talking loudly and raising dust with every dribble.

On the back porch of Glenda's home, Antony sat down on an old couch, just watching the soccer game that brought him back childhood memories.

On the way to her home, Antony had told Glenda the whole story about his book and the Angels who wanted to destroy Cloud Base. Glenda loved this idea because in the past, she and her husband, Allan Jackson, had gotten into a bad deal with the Sims Family.

Ten years ago, Glenda and Allan Jackson created *The Battle of Iron Giants*. They had always worked with robots in civil constructions, the maintenance of roads, hydropower plants, and junkyards until, one day, they created this kind of sport on a wasteland where the community was located. The news of this kind of fight spread quickly, drawing the attention of other robot pilots in America and a huge crowd who loved to support and bet on their favorite robots. The fights were getting increasingly famous, making a TV producer full of ideas called Toby Maison sign an agreement with the Jackson couple to broadcast the fights on TV and the Internet. The success was immediate and absolute. Glenda and her husband left the fights to focus only on the management of the community around the arena and the fight organization. They started buying lands and building houses near the arena — it was the beginning of the community. Also, many illegal immigrants and poor people from other states were attracted to the place in pursuit of a better life. Glenda and Allan financed houses and gave jobs to the newly arrived in the community. Toby Maison changed the name of the show to *The Maximus Arena*, following Allan Jackson's idea — a name that sounded more powerful, according to the patriarch of the community. Everything was going perfect, and only promising opportunities seemed to be ahead of the Jackson couple and their new land... until a man called Clifford Sims offered good money for Glenda's enterprise. Glenda told him that her business wasn't for sale and that most of the land already belonged to her ancestors. Clifford Sims was one of Antony's cousins — someone who controlled everything related to media around the world and hated when someone didn't accept his money.

A few days later, Clifford bought all the Internet companies and TV stations that broadcast the fights organized by the Jackson couple. The audience fell to zero, the money became insufficient to pay their taxes, bills, and the wages of so many immigrants and employees. The situation got worse after Allan Jackson and his oldest son, Thomas, passed away in an accident involving one of the *Dockbots*. Glenda was so desperate that she decided to look for Clifford Sims to sell her enterprise, but it was too late. According to Clifford, *The Maximus Arena* lost its market price and audience — so, it wasn't worth as much as it had been a few months ago. *But you were the one to blame for the show losing its audience, Clifford,* Glenda said, even though he refused to make business for a while.

Things went from bad to worse for Glenda, who fell into a deep depression, especially after some men to whom she owed money threatened to kill her. *Ok, if I kill myself, I'll go to that damn Cloud. If they kill me, I'll go to that damn Cloud too*, Glenda kept thinking this way for a long time until Clifford came back to buy her show, but for a low price. She didn't have any other option except for accepting Clifford's proposal, so she could at least pay the men she was in debt with.

After Clifford bought the rights to *The Maximus Arena*, Glenda thought, *Well... now he has the rights to the show, and he'll make a lot of money managing the fights*. However, that didn't happen. Antony's cousin bought the show but didn't broadcast it anymore. He claimed this show incited violence, bullying, and a rude way to solve the problems of life. Glenda didn't understand how someone could spend so much money to buy a show just to end it. *Too much money can fry the human brain*, she deduced.

After overcoming all these troubles, Glenda Jackson lived for her surviving son, William, and for the rest of her community who remained there — day by day living in a simple and humble way. Sometimes she organized a fight just so the people there had a reason to smile and cheer. But she couldn't broadcast any fights on the internet, or else she would have to pay an expensive tax to the Sims Family, according to the agreement she had made with Clifford.

CHAPTER 32

"The idea for *The Maximus Arena* was your husband's. That's insane. I always thought it had been Toby Maison's."

"Allan didn't like to be in the public eye, and he especially didn't like to be called by his first name. He preferred to be called Mr. Jackson," Glenda spoke sadly.

"I'm very sorry... I know you hate my family and Cloud too. For you, it's a tragic coincidence my grandad and your husband had the same name, I guess."

"Yes... I can't remember the man I loved without remembering the one I hate so much. Sorry bro, this damn Cloud is destroying the world."

"Yes, I know."

"You know? And you want to help save this place. Gosh, you playboys are crazy. So, why do you work with your family?"

"I was born in it. Cloud is a reality that has been shown to me since I was a child. I hate that place as much as you, but I can't simply destroy it. If I did, my family would just build another one and another one... we can't destroy the place... we have to destroy the idea that everybody depends on it. Now these Angels are showing us that this place is dangerous, and that we need to rethink the idea of Cloud."

Both stayed silent for a while, as if they were expecting the silence to bring a solution to this tense meeting. They sat down on the old couch, on the back porch of Glenda's home, watching the guys play soccer on the sand field ahead. Each one drank an iron mug of black coffee.

"You have probably seen the video of the two Angels fighting in the Cloud Base," Antony broke the silence, "more of them might come here and destroy this damn place. But my greatest doubt is 'what do they want?' For me, it's the beginning of a new kind of war, worse than a battle between humans and aliens in sci-fi movies or the ancient World Wars that we studied in school."

"Have you thought about... you know... turning Cloud off? Like the Angels want."

"Yes, I have. But like I just told you, my family would just turn it back on. Also, believe me, there are good people there, and they are happy."

"Happy? For me, it's a strange world, bro."

"What do you mean?"

"Some years ago, my husband and my oldest son, Thomas, were working on the maintenance of a robot in the big *Dockbot* next to where you landed your toy plane," Glenda looked angry, her eyes watering. "Your damned cousin had already ended our business with the fight show, so we had to accept whatever job came to us to make some money... It was about midnight, and we were working tired... very tired.

Suddenly, a heavy iron bar fell from the roof of the *Dockbot* and hit my men hard… The death was instantaneous."

"Oh… I'm so sorry."

"When I talked to Cloud's attendants, they told me that there hadn't been time enough to transfer them to Cloud. The accident happened too fast, according to them. I didn't know if I felt sad for their deaths or happy for not having to see them in this damn heaven created by your grandfather. Now… I live for the people who remained here and for my son. Sometimes, I'm afraid of dying and becoming a 'prisoner' in Cloud. These damn chips they put in our brains… why? Do you know how to take it off?"

"They… they have it installed in our brains when we're newborns. Over time, our brain develops and covers the chip with a kind of membrane and cephalic mass," Antony looked sad, "it's impossible to take it off."

"Well, at least my men had the opportunity to see God. It comforts me."

Antony noticed a little crucifix in one of her silver bracelets, a symbol that had almost vanished but that immediately made him remember his mother.

"Nice bracelets. Nowadays it's very uncommon to hear someone talk about God."

"Well, it's better we get to the point," Glenda gulped down her coffee. "You want me and the other guys to protect Cloud Base with our robots. Just because you imagine the next Angels might be giants?"

"I'm not just imagining; the Angel called Muriel told me. I don't know how many might come, but we have to be prepared for that."

"You've already told me about this Angel. So, what makes you think that I will accept to help you? You know I hate your family and Cloud, bro."

"If you had the opportunity to prove your worth to the family that almost destroyed your life and, with that, help your son and your community, would you do it?"

"And do you think that is the way?"

"What other opportunity would you have to approach my family? I'm not the owner of Cloud... I'm just known as the rebel grandson of Allan Sims. The other members of my family don't listen to me much. Now, they are feeling threatened. That's our chance to end this monopoly," Antony was excited but serious.

"What about your damn cousin Clifford? Is he still destroying other people's lives?"

"He is the pacifist of the family, always with his hippy ideas... Love and peace... you know. It's damn insane because when we were kids, he was the worst troublemaker. I still remember he called me *abortion leftover*; I never wanted to know the reason."

"'Abortion leftover', oh... that is a damn disgusting nickname. If I were you, I wouldn't even want to know the reason either."

"I agree... Anyway. When Clifford became an adult, he enjoyed his position as a rich heir just traveling and practicing meditation in the middle of nature. Probably in one of these travels, he fell from a mountain and died. Because now he's living in Cloud doing the same shits he did here."

"I don't understand you, playboys. You have enough money to do nothing, but out of nowhere you decide to control the behavior of everybody. Dictating what people have to do or not."

As Glenda talked and watched the guys happily playing soccer, she had an insane idea.

"Ok, bro, what do you think about a soccer match with the boys there?"

"A soccer match?" Antony looked astonished and doubtful at same time.

"Of course. If you score at least one goal, I'll help you," Glenda let a sarcastic smile escape from her mouth.

"Are you serious? Do you want to decide the destiny of the world with a soccer match?"

"Your crazy cousin ended my robots' fights claiming that I was practicing bullying. Now I'll do like for like, bro. Twenty minutes of a soccer match to define the destiny of the world. Take it or leave it."

Antony knew very well that he wasn't in his territory, and he couldn't solve it through some commands on his phone. He had to accept this odd proposal to keep his plans going.

"All right, all right... I don't see how this makes sense, but I accept this insane challenge."

Both got up, leaving their mugs on a table next to the sofa. Glenda called everybody in an imposing voice and explained the challenge to the guys. She didn't say anything about the Angels or Cloud; she just said that the visitor needed some help, and he should score at least one goal to deserve it. There were all kinds of guys on the field: thin, fat, African Americans, Asians, and Latinos from fourteen to seventeen years old. The younger ones and the girls, who were between eleven and thirteen years old, stayed around the field supporting the teams.

While the guys were preparing the teams, each with six players, Glenda secretly whispered to William and the others from his team, *Go rough on the playboy.* Her son just winked, and the other guys nodded discreetly.

William was a strong boy who looked like his mother. Always ready to obey Glenda, especially on the hard work in the *Dockbots* that he loved to do. His head was shaved, and he often wore a mechanic's uniform, except when he was on the sand field, where all the boys were barefoot, wearing T-shirts and shorts.

"Hey, playboy, have you ever played soccer in your life? It's better you take off your high heels before starting, hahaha," William mocked him and laughed.

"Yes, I have, when I was a kid. And you're right. It's better I take off my… shoes first," Antony looked suspicious.

He slowly took off his shoes and folded up the legs of his pants. Also, he realized William was on the opposite team, and Glenda, minutes before, had whispered something to him. "These guys are troublemakers, I'm sure," Antony thought. He was not so good at soccer, but he excelled at calculus and reading the moves of the games.

A whistle sounded from the mouth of a little boy on the corner of the field as Glenda just watched with crossed arms.

"Hi, guys, only twenty minutes go, get ready," Glenda shouted out.

Several little children were around the field and close to Glenda, supporting the team that William was playing on.

The match kicked off, and Antony's team had the ball. Kids cheered, tension filled the air, and the smell of sand dominated the scenery.

"Gooo, William," the kids yelled.

Antony got the ball for the first time and advanced fast, but, in his first move, William kicked his legs hard, making him fall face-first and stay down for a few seconds, in a lot of pain.

"Woooowwww," the kids yelled.

William walked near him and kicked some sand onto his face, making the kids laugh harder.

"You'd better bring you're A game, playboy." William said.

"I think so, aaargh," Antony replied.

The match restarted, and so did the aggression towards Antony. In the first five minutes, Antony was knocked down seven times. He was covered in dirt like a miner on duty days, and his legs full of bruises.

William's team scored the first goal, making the kids go crazy. Two minutes later, William's team scored one more goal and knocked Antony down three more times. It was evident that William's team was stronger than Antony's, and Glenda had made it so just to humiliate her visitor and dismiss him.

Antony was sore all over when he approached Glenda, who said sarcastically:

"Hard match, bro? These guys play for real."

"Yes, I noticed that. Aaargh... what if we double the bet?"

"WHAT?" Glenda looked startled.

"I score a goal and my team wins. What do think?"

"Bro, you're crazier than your crazy cousin. I'll love to see it. I agree," Glenda finally smiled.

Antony went back to the field and called the guys from his team to him.

"Hey guys, come here, please. Let's plan our next steps."

Everybody from Antony's team was a little down, but they got together in a circle around him, while the other team just relaxed and told jokes on the other side of the field.

Antony changed his mood and looked confident, pointing at each guy he was instructing.

"Ok, guys. I know you take all matches seriously, and you'd love to kick their asses. Am I right?"

Everyone nodded.

"So, this is the plan. I'll run to the right to attract William. He can knock me down, but he's right-handed. It will be difficult for him to kick to the goal. You go to the left and draw the other tall guy who is southpaw and hates you and just wants to kick you. I don't know why."

"I stole his girlfriend," the guy looked goofy as he replied.

"Wow, we have a lady-killer on our team, that's insane," Antony continued. "Ok... These two will have some difficulty kicking to the goal. We'll get the fat one, who is flustered, to go forward and try to score the goal, but he's very bad. This is the opportunity for you to go fast in the middle because the way to the goal will be open. Meanwhile, you have to stay close to the goal... You... You're a good defender, keep your focus on the fat one and the Chinese one. Leave William to me. He's strong but he's slow for me; it will be easy to escape him and pass the ball to you. Ok, guys, that's the recipe for success."

"Where did you learn so much about soccer?" A boy asked, doubtful.

"Wow... in statistics class. Now you have to go," Antony looked confused as he answered.

He was feeling something in his chest that made his face show pain.

"Something wrong, bro?" another boy asked.

"No worries. I'm ok. Now... here we go!"

The match restarted, and the Antony's strategy showed good results. William's team was lost in the match. They approached the goal, but they were unable to finish an attack.

Antony quickly dribbled past William and passed the ball to another guy on his team who scored a great goal.

"YES... WHO IS THE BOSS, GIRLS?" The guy who scored the goal shouted out.

The match continued, and Antony's team dominated all the moves until they scored another goal. The match was tied.

"GOOO, PLAYBOY," the kids yelled, making Glenda annoyed and nervous. She never thought that a rich guy would be able to organize a team of poor teenagers.

There were only seven minutes left for Antony to score his goal.

The guys played bravely, and when there were four minutes left until the end of the match, Antony's team scored another goal. Everyone cheered a lot, and William's team became increasingly angry and lost focus on the match.

Despite Antony's team winning, he still needed to score his goal, according to his agreement with Glenda.

William was so nervous about the result that he couldn't stop Antony's quick dribbles.

"Hey, bro, how did you become so good in such a short time?" William asked.

"Ask your mother, bro," Antony shouted back.

The kids laughed a lot, and everyone was supporting Antony's team. William was so worried about obeying his mother that he lost all focus on the game and just continued all the time in pursuit of Antony, who was faster than him. Antony wasn't a good player, but William didn't realize that he was on the wrong side of the field, where it was difficult for him to possess the ball.

Only two minutes left until the end of match, and Antony's heart started beating fast, but he was so excited with the way the game was going that he didn't even realize it.

"Come on, bros. Let's finish this game once and for all," Antony screamed."

In the middle of the field, a guy from Antony's team ran, possessing the ball. Then, he passed it with a light kick to the goal area. Antony ran ahead and dominated the ball, going with it toward the goal as William pursued him like a furious lion. Only one minute left, and there was nobody between Antony and the goal, only a few steps ahead. Glenda's face showed disbelief as she saw her visitor advancing through the field. Antony smiled as he approached the goal, making him, for a short moment, remember good childhood moments when he played soccer with his cousins — without worries. But suddenly, something unexpected happened.

For a moment, Antony's heart stopped, and he felt something cold in the middle of his chest. His eyes went totally white, and an intense weakness took over his body. His last movement was a weak kick on the ball, which went slowly toward the goal.

Antony fainted next to the goal and fell hard on the ground, raising a lot of dust. Everyone thought he had just stumbled, but when all of them realized that he wasn't moving, a desperate feeling spread through the field.

"He's dead! God, he's dead," Glenda's red lips were trembling as she whispered to herself.

Everyone was scared, watching a dirty guy lying down on the ground, seeming to be dead.

CHAPTER 33

The hands on the clock on the wall of Christine's office were approaching 1:00 p.m. as she took a big, green, tattered suitcase from a secret compartment below her desk. This was an old telephone that used a radio signal, like the one used by Uncle Peter, who was waiting to talk to his niece in California. Also, there was another important person who had this kind of equipment.

Christine kept her eyes on the clock until it showed 1:00 p.m. Then she took the telephone from the suitcase. The connection between the three people wasn't great; there were some crackling noises from time to time, but it was good enough to talk through.

"Hello, are you two there?"

"Yes, Uncle Peter speaking. How about our guest?"

"Let's wait a little more; this kind of telephone doesn't work well over long distances."

"Hmmm, and how about your loving cousin?"

"Your loving nephew... He went out, with no warning. I was trying to contact him but no replies. Also, I tried to locate his jet, but it takes time."

"Take it easy, Chris. Sometimes you should admit that he is smarter than you, and if he doesn't want be found, he can do it easily."

"No, I can't take it easy. He worked on and developed the security system of Cloud, and he added nothing to the last meeting," Christine sounded angry.

"Maybe he knows this system so well that he decided to pursue another solution."

"I agree with you, Mr. Peter Sims," the voice of Yelena Orlyk rang through their phones.

"Good to hear you, Prime Minister," Christine's voice sounded a little strained and shy from this moment on.

"I didn't have a good experience at the beginning of our last meeting. Well, does anyone know Antony's whereabouts?"

"He probably went looking for a solution to our problem," Peter led the conversation. "He has always acted this way. You know, Chris, he thinks better when he acts alone."

"I don't know. I have my doubts about what side he is on."

"He's part of your family. There are no better people to know what his next steps will be."

"I told you, girls. He is a boy who craves the love of his family. Certainly, you've noticed that he has changed his behavior. If we treat him well, we'll have him on our hands. And he'll solve all our problems."

"You're right, Mr. Sims. I realized that he was more... I could say... accessible on our last meeting. But for me, a person who lets themselves be driven by fake family feelings is not a very strong person. I hope our little scene with Mr. Antony and the president resulted in a solution to our problems. His personality is weak, sometimes explosive, and sometimes confused. But I have to admit that he's a genius when it comes to machines and strategies. That's all we need from him."

"Moreover, we have to pray he doesn't use Key. If he decides not to use it, we have him in our hands."

"You still insist on this fake story about Key, Uncle Peter. Antony never admitted to it."

"And he never denied it either."

"Christine already told me about this supposed system that you call Key. For me, it sounds like such a stupid idea to create a simple off button for Cloud. But from Antony's mind, we can expect anything. Keep your eyes on him. Let me know whenever there is any new information."

"Ok."

"No problem."

They turned off their telephones at the same time.

CHAPTER 34

"Hey, bro, look at the scar on his chest. It's enormous."

"Yes. He had surgery on his lungs, I think."

"Maybe. This scar is bigger than my dick."

"Anything is bigger than your dick, bro. Hahaha."

"Shut up, motherfucker. I think he's waking up."

"Get your fat asses out of here, boys. RIGHT NOW! Glenda yelled at them.

Antony slowly opened his eyes. He was lying down on a couch in Glenda's living room as William and another guy who had been playing soccer with them were attending to him. The rest of the guys who were in the match were outside watching the sick visitor from the windows.

The living room was a little messy, with two more old armchairs and a wooden coffee table where Antony's nylon jacket lay. There were two massive wooden shelves with several books, family portraits, and high school awards. On one of the walls, there were four flags of African countries, and in the middle of them, a metal crucifix stood out.

After kicking the boys out of the living room, Glenda crouched down beside Antony, who moved slowly to sit up on the couch, still feeling weak and lost. He was not wearing his nylon jacket, just his tank top.

"Hey, bro. Are you ok? Please, say yes." Glenda looked worried.

"Did I score the goal?"

"No."

"So, I'm not ok. Oh… my damn head hurts so much."

"Hey, bro, I'm serious. Please, don't joke with me."

"Don't worry. I'm ok… I forgot that could happen again. This is what happens if I push my 'strong physique' too hard," Antony spoke calmly.

"Has this happened to you before? When I took off your coat and saw this big scar… oh, God… I thought I'd killed you. You should have told me about it."

"And would you have listened? Don't be anxious; it was only a few minutes."

"A few minutes? You were unconscious for more than an hour."

"Wow… that's never happened before," Antony looked surprised.

Finally, Glenda looked calmer when she sat down beside him and told the guys to leave the house and walk away from the windows. Antony saw some of the guys who had been on his team.

"Ok, bros, our strange visitor is doing just fine as you can see. So, get out of here and let some fresh air in the house."

"Yes, boys. I'm alright. Hey, you were fantastic. You deserve to go to the World Cup, if it still existed, of course," Antony pointed at some guys at the window.

"You're fantastic, brother. I hope to see you again next match. Stay safe," a guy said as he walked away from the window.

"Ok, time to explain what happened on the field," Glenda said.

"Well… I was born with a weak heart," Antony felt unwell, but he smiled shyly. "There is no cure. So, after many different treatments, the doctors decided to implant a system around my heart with valves and a high-tech pacemaker with hmmm… a kind of artificial intelligence."

"What? That's crazy, bro. I've never heard of anything like this."

"I'm the only one who has this kind of system. First, because I have the means to hire the best doctors to develop this system, and second, because it's very unlikely anyone else in the world has a heart as weak as mine."

"But this system is not good enough, I guess. You couldn't stand a soccer match."

"You're right. My heart still can't handle much effort. But during the match... oh my head... This system decided to make me pass out until my heart could go back to the normal rhythm."

"Oh... that's incredible. At least your chances of having a heart attack are minimal, I guess."

"Now, I don't know. If there's any risk to my heart, the AI identifies it, stops it for a moment, and then the system... you know... restarts it later. But an hour? Damn. That should last two minutes tops."

"But you are alive, bro, and that's what really matters," Glenda breathed relieved. "I'd hate to be accused of killing the heir of the most powerful family in the world. I think I'd be arrested in a jail inside Cloud. If I got lucky, of course."

"You are lucky. Your son and friends are amazing. Believe me. Being around good people is a blessing, and there's no amount of money that can buy that."

Glenda got emotional and quickly wiped a tear from the corner of her eye.

"Well, it's time to go," Antony was ready to get up, "the agreement was that I had to score a goal, and I lost the opportunity."

"Hey, bro, take it easy," Glenda stopped Antony from getting up with a touch on his shoulder and started talking calmly, almost like an apology.

"When you were unconscious, of course, I couldn't stop thinking that I killed you. God... I remember you told me there were good people living in Cloud. And... and you asked me to help you protect them. So, I felt horrible just thinking I had been responsible for your death, imagine being responsible for the death of millions of people."

"I don't wanna put the responsibility of Cloud's population on your shoulders."

"But you came to me. And for me, well… I know you don't believe that, but for me, it's God's call."

"I believe that more than you imagine. My mother is a religious woman, and she taught me a lot of important things… So, about my request for help. Have you thought about it?"

"Yes, I did… well… I can try to help you."

"Really. It's amazing," Antony looked surprised and excited.

"For a price, of course. You're talking to the best robot fighter in the world," Glenda looked more friendly and relaxed. "Also, you got a victory for your team. That was crazy, bro… and fun too. Except for the dramatic ending, of course."

"Hahaha… yes, it was insane and great at the same time."

Antony smiled and looked curiously at the flags and the crucifix on the wall of the living room.

"Also, I don't believe that saving the lives of Cloud is your only reason to help me. Sorry, sometimes I have strange presentiments about others. Can you explain what really made you change your mind? If you don't mind, please."

Glenda took a deep breath and started to explain.

"Your presentiments are incredible, bro. Well, can you identify these flags on the wall?

"Hmmm… Angola, Nigeria, and —"

"Mozambique and Ghana," Glenda continued. "My ancestors probably came from some of these countries, according to my grandparents, who were told that by their grandparents. Anyway... before my ancestors were taken to America, they lived happily in Africa, but for ages, they fought against other people. People from different cultures but with the same skin color. When the fight ended, the victorious people enslaved the losers and sold them to Europeans who brought them to America. I could say... black people enslaved black people to sell them to white people. Everybody hates me when I say it, but in truth, it was never a matter of racism; it was a matter of power and ego."

"I had never thought of it this way. I'm so sorry."

"No problem, bro. In the past, people were taught to not like others who have my skin color by prejudice. And years later, black people were taught to hate white people to protect political ideologies or something. In the end, all this shit education did was hide the real reasons."

"Power and ego. It's so clear."

"Exactly, and I agree with you. I would love to see Cloud destroyed. But if we destroy this place, other rich men will come and will rebuild it. We can't destroy only the place... we must destroy the idea. And to destroy the idea, we must show our worth to these Angels and to the rest of your family and to these damn politicians."

"Great. My family's Corp and its henchmen in government are very powerful. We can't treat them like we're the rebels; we have to work together with them for a while and win over their hearts and minds."

"Yes, it doesn't matter if the community is living a simple life. I want to see everybody free. Free to think what they want. Free to believe in any religion and face death however they want. Like you just told me, they are amazing people, and I'd hate to lose them. I don't know why, but I have an odd feeling about the survival of these folks."

"What do you mean?"

Glenda pointed to the window at the teenagers and kids, boys and girls, who were playing soccer outside, not seriously like the last match, just having fun.

"Can you see the smallest one? He's Joseph, the last kid who was born here. He's ten years old. In ten years, nobody has been born here. And one thing that these crazy guys love to do is trying to make babies, if you understand me. I have read weird theories on the internet that say nobody else will be born anymore and the people will go extinct. God, it's hard to believe that, but I confess I'm very afraid. Have you heard about that?"

"I have read about it, but I don't believe it. The world has been so weird in these last decades, and these Angels are just proof of it."

Antony was impressed with himself for sounding so cold in his last answer. He was unable to confess that he knew the truth. It would be hard to choose the words to describe that little by little this community and the rest of humanity were vanishing. A deep agony stabbed his heart like a blade, so hard that he thought for a moment that his pacemaker would knock him down again, but it didn't happen. Furthermore, these last words from Glenda made Antony feel even more responsible for these people. *If you have power enough to make the world less bad, just do it,* this was his favorite quote from his mother which he was remembering in that moment — making his shoulder heavier with the weight of the world.

He couldn't find words to keep the conversation going, but luckily Glenda changed the subject and started to talk about the Angels again.

"And the hearts and minds of the Angels, bro? Do you think it's possible to change those too?"

"Yes... the Angels," Antony recovered his breath, "I noticed that these Angels have a kind of ethical code. If one wins a battle, the other one respects it and doesn't seek revenge; they don't break the rules."

"And if we win a battle against them, they will respect us."

"I hope so."

"And we'll be able to withstand their strikes? Because bro... I watched the video of the two Angels fighting inside Cloud Base, and I think they're bulletproof. And their armors seem to be from another planet."

"It would take missiles to shoot them down them like it happened in Paris some days ago, and Cloud Base is well equipped with these military weapons. But anyway, we have to be ready in case these weapons aren't enough, which is what I think will probably happen."

Antony took the piece of Muriel's armor from one of his jacket's pockets that was on the coffee table. Then he showed it to Glenda.

"This is a piece of the armor that belonged to one of the Angels you saw in the video. One of these Dockbots probably has a Laser Fibers Lathe to create different kinds of metal. If you don't have the elements to create it, I can get them easily. We can make armors and big medieval weapons to fight against these Angels. We can fight like them, fight with the same weapons. This way we can gain their respect."

Glenda took the solid piece of metal and started to analyze it carefully. Its surface was smooth, and its color was gray, covered in blue lines that seemed to shine.

The strong woman seemed to fly through her own thoughts, going through all possibilities and chances to protect those she loved. She knew very well that she had to make a decision in that moment and that decision would certainly change her life. After a brief meditation, a combination of instinct and rationality brought her the best answer.

"Ok, bro, time to kick those Angels' asses and the rest of your family with just one big kick."

Glenda and Antony looked very proud when they shook hands firmly, asserting a strong friendship.

CHAPTER 35

Glenda, her son, and Antony were walking toward the biggest Dockbot of the community through a sand path with some bushes surrounding it.

William was silent and looked awkward as he walked. He was wearing clothes similar to his mother's, worn-out denim overalls and a dirty white long-sleeved T-shirt. Suddenly, this big and strong guy gained courage and talked to Antony.

"Hey, bro, sorry for kicking you so many times during the match. You know... it's part of the game."

"No worries, man," Antony smiled, "you were just obeying your mother, you're a good son. And a good son always obeys his mother, even if what she told him to do is a big mistake."

"Yes, bro, you're right," William looked more relaxed, "my mother asked me to kill you and I almost did it."

"Hey, watch your mouth, boy. It was just to scare him."

"And you got it. Hahaha..."

Glenda slapped William's head as Antony laughed.

"Hey, William, your mother told me that you're one of the best robot fighters in the community. Is she right?"

"Of course, bro," William boasted, "I'm able to knock down anybody here. Even my mother, if she doesn't use the *air kick*. It's impossible to defend yourself from it."

"Unfortunately, he's right. Well, I have trained William since he was a kid with little robots. When he got tall enough to pilot the big ones, he already knew all the tricks to win."

"You were probably training him for *The Maximus Arena*."

"Yes, bro... oh, bad memories again."

"Oh... sorry. I couldn't resist —"

"When I was a kid, I dreamed of fighting in the arena every single day," William looked excited. "But now, thanks to Antony, I can show my talent against these damn Angels."

"Are you crazy?" Glenda looked angry. "I'd never let you go on this suicide mission."

"Come on, Mom. You just said I'm one of the best fighters here. Our new friend, Antony, needs all the help he can get to save the world. So, we kick some Angels' asses, then Antony can teach me soccer dribbles."

"Me? Oh, man... I'm a horrible soccer player. I just used the best strategy at the right moment. If you kick with your right leg, you can't stay on the right side of the field."

"What do you mean, bro?"

"Humm, ok... let me explain."

William and Antony got themselves into a deep conversation about soccer as they were walking toward the Dockbot. Glenda stayed silent, just paying attention to her only son, finally having a good talk with someone different — and she was loving it. People in this community didn't have the opportunity to study, and William was the best example of it. A guy full of energy who was always smiling and joking with his mother. He could learn anything fast and would have had a promising future in The Maximus Arena. However, he lost his father and older brother in a tragic accident. After the show was canceled, William and Glenda focused their lives on working hard to pay their taxes and help the community. In the golden years, Glenda had planned to build some schools, courses, or even a college here with the profit from her enterprise, but things hadn't gone as she expected.

Antony showing up at the community seemed strange and dangerous to Glenda in the beginning, but in that moment, she realized that this guy brought new perspectives that had already vanished from her mind long ago. For her, it didn't matter where Antony was from. He proved he was honest, and if he could help her only son and her community, she would go with him until the end.

They were approaching the open gates of the big Dockbot that looked like a huge medieval castle made out of metal with large gears around it. There were seven giant robots, each about a hundred feet tall, made out of pieces of different vehicles like cars, trucks, airplanes, and tractors. Antony noticed that some of them were very rusty and there were bushes growing around their feet.

Inside the Dockbot, there were many big, heavy machines to build these giants — long mechanic arms fixed on rails from the roof, bigger than tractor shovels, held big tools like drills, torches, robotic claws, and sprays that squirted oil and water. In the back of the place, there were ten tall metal silos, almost the same height as the Dockbot's interior, some with heavy and powerful furnaces ready to melt any kind of metal and spread it in large flexible iron molds.

To the right, there was a metal cabin with many black keyboards and four wide transparent screens. In front of them, a metal support, that looked like a two feet high stool, with a metal claw above it, was ready to work. Several electric cables connected the desk with this support, and another machine beside it looked like a truck engine with a *gun* pointed directly at the metal claw. This was part of the *Laser Fibers Lathe*, a system comprised of several machines able to identify the minerals that composed any kind of metal and improve or replicate it. The point of the gun touched the piece of metal fixed on the metal claw with a laser beam, then all information about the composition of the metal was sent to the silos, which mixed all kinds of molten metal and spread the mix into molds.

In the middle of the Dockbot was a new robot that was still missing some parts of its chest and arms — even though it looked powerful and imposing.

"That's an insane work-site; I never imagined that my favorite robots were built here," Antony looked mesmerized. "This robot is amazing. Are all the machines here working? And the computer system?"

"Yes, everything works very well here," Glenda answered calmly. "The place is a little dirty and messy, but... it's our place. This is a pretty robot I have been working on for the last few months. Just for fun, to be honest."

"Come on, Mom. You hope The Maximus Arena comes back."

"That is the wish of everybody here in the community, I guess," Glenda's eyes sparkled.

"Well, can I use your Dockbot's system, please? I have to know what this place can to do."

"Sure, just press any button on the panel, and it will all turn on. Do you know how to use it?"

"Of course, it's like a video game."

Antony looked excited as he fixed the piece of Muriel's armor on the metal claw. Then, he positioned himself in front of the transparent screens and keyboards. Just pressing one random button on a keyboard made Antony's curious eyes look attentively around, waiting until everything was turned on.

"Does he know what he's doing?" William whispered to his mother.

Suddenly, Antony started typing on the keyboards at an incredible speed, keeping his eyes completely focused on the screens, which showed information about the Dockbot on several little images that appeared quickly, one over the other. Also, many algorithms fell from the top of the screens like a waterfall.

Glenda had never seen these little screens and algorithms while she worked there.

A laser beam hit the piece of armor, which shone red as the system sent more and more information about the strange metal to the computer. The Dockbot's interior lighting, coupled with the lights coming from the screens, made Antony look like a dummy covered in green and red numbers and words. All the Dockbot's machines started moving slowly and buzzed like the whole place was an iron cathedral, and Antony was a phantasmagoric pianist playing an organ.

William and Glenda were mesmerized as Antony didn't stop typing, focused on the screens. For a moment, mother and son thought that their visitor was hypnotized, but he talked normally as he stayed in action.

"Well, this metal has a kind of mineral combination that I have never used, but it's easy to make. However, we'll need more minerals like gold, lead, niobium, zirconium, gadolinium, and, humm... aluminum. It'll be an interesting mix, and I hope it becomes a resistant armor. Your system is working very well, Glenda."

"Hey, bro, I forgot to warn you that the Internet signal here is not so good."

"No worries. I've just expanded the signal ten times; that's enough to work. Oh, I almost forgot to pay you for your great help. Let me do it right now."

Glenda heard her phone buzz inside her pocket. When she checked the message from the bank, she was hit with a big surprise. A huge amount of money had been transferred to her bank account. She had never seen so much money in her life, even in the times of The Maximus Arena. William, standing by her side, saw it too and was hit with the same surprise.

"Just remember, this is half of the money. I'm gonna send you the rest after we... you know... finish the job."

"Bro, for this much money, I'd kick even demons' asses. That's crazy!"

"Watch your mouth again, boy. Hey, Antony, what about the minerals that are missing?"

"Besides that, I still need some other things you don't have; reactors, titanium screws, suspensions hoses, some carbon slabs," Antony kept typing. "Don't worry, I just asked the Sims factory close to Kansas City. All this stuff will get here in... humm... about three hours. Do you know how many robots we can equip in four days?"

"If we work fast, three of them, I guess."

"Well, this is all we can do. Now, we just have to wait for the equipment."

Finally, Antony stopped typing and relaxed, which made mother and son relax too.

"Hey, bro, I thought you'd never stop," William was still amazed, "but after the robots are ready, how can we go to… to the place? The Cloud damn Base."

"There are train tracks that start in Kansas and pass nearby, about thirty minutes walking by robots. Then I can combine the train tracks until we get the one that take us to Cloud Base. That's the plan for now. We'll equip the robots, wait for the train, and then it will take us to Cloud Base. I hope we have enough time."

"And how will this train take us?"

"I can control it."

"Are you a train driver, bro?"

"No, but I have a phone."

William looked doubtful but nodded.

***

"Ok, ok, Christine, don't worry. I know very well what I'm doing."

"I believe you, Antony, but robot fighters from an old TV show? That sounds ridiculous."

Antony was alone, sitting down on a big tire surrounded by some bushes close to the Dockbot. It was evening, the sky was orange, and the sun shone its last rays on Antony, who was talking to his cousin on the phone.

"A fight between two Angels sounds ridiculous too. But it happened in your workplace."

"You're right. I admit it, but for me, it still sounds ridiculous. We are equipped here with more missiles that we can imagine. Remember Paris. A drone knocked your *winged friend* down with just one missile. Imagine what can be done now that we have hundreds of them."

"Muriel let herself be hit by that. Don't belittle her. Besides, she's not my friend."

"All right, but she's still here, like a dying patient just looking up at the roof of the infirmary."

"It's better we keep a close eye on her."

A craft flying over in the sky drew Antony's attention.

"Ok, Christine, my order is arriving. I have to hang up."

"No worries, keep safe, dear cousin. See you soon."

"Dear cousin?" Antony whispered to himself after Christine hung up.

Antony could hardly remember the last time his cousin had treated him so carefully; however, for him, things were becoming better despite all the challenges.

Three hours after Antony ordered the materials from one of the Sims' Factories located in Kansas City, a white KC-490 millennium, a medium cargo plane with retractable turbines, landed between the Dockbot and the blue Bombardier Global 1K. From this aircraft, some beige androids unloaded all the necessary stuff to equip the giant robots on small-wheeled containers the same size as a refrigerator: bars of different kinds of metal, screws, suspensions hoses, some carbon slabs, and many heavy weapons like small missiles and machine guns. There weren't any human pilots or employees on this aircraft, just the AI and androids.

The kids of the community, as well as some adults, admired the androids unloading the stuff with curious eyes, like a scene taken out of a sci-fi movie.

Antony and Glenda were organizing everything inside the big Dockbot when William arrived with seven more hard workers. This team was a faithful representation of the ethnic mix of the community: Americans, African Americans, Asians, and Latinos.

"Hey, bro. These are the best metallurgists of the State: Joe, Malcon, Selena, Robson, Jane, Haitao, and Latoria. They were born inside molten metal, if you know what I mean."

"Nice to meet you, guys. Well, we can start equipping *Big Glenda* here. It's the newest robot, and we can see some fast results of our work."

"*Big Glenda*? Oh no, bro," Glenda disagreed, smiling. "I hate giving names to robots. I'd rather call them by the pilot's name."

"Like I said. Big Glenda hehehe... Here we go, guys."

"Come on, bros... move those fat asses."

The night was just beginning when the white KC-490 took off, and everybody inside the Dockbot started to work. Smoke, long robotic arms, steam, gears turning, tubes spreading molten metal into huge molds, and water being poured on it. Humans and machines worked like one huge sole organism that smelled like smoke and metal. Antony kept his hands on the keyboards, and his eyes were everywhere. The metallurgists acted bravely and with pleasure, turning levers and managing all gears with great ability.

Over the night, members of the community came to the Dockbot, bringing meals and drinks for the workers who got it in short snack breaks. Antony had never seen this kind of behavior among people working in the same place. There weren't any supervisors, managers, or quality inspectors to check on the employees. Everyone knew what should be done in complete harmony.

"Hey, Glenda, come here, please," Antony called.

"What's the matter, bro?"

"Do you think this size is good for you?"

Antony showed Glenda a mold that he designed for her on one of the screens.

"A sword? Bro, you're damn crazy. Is this for me?"

"Yes. It'll be our first weapon made from the same metal as the piece of Angel's armor. It looks like it's really resistant and soft. Here we go!"

A huge structure, that looked like a long pool surrounded by several thin metal cylinders fixed on a kind of train tracks on the floor, formed a flexible mold controlled by robotic arms. This setup made it possible to create several shapes out of large pieces of metal. Through the system, Antony designed the shape of a medieval sword that was fifty feet long, half the height of the robot. The silos poured incandescent molten metal, resembling the surface of the sun, into the mold. Then, long robotic arms poured water over it, while as other arms holding big sledgehammers hit the new weapon several times. The movements of the metal arms, the clanking of metal, and the steam that arose from the mold made everyone feel like they were inside a huge running engine.

Antony turned off the thin cylinders off on the computer controls, revealing the first finished weapon. A long gray medieval sword with some blue tones that shone on the reflective blade.

Glenda approached this weapon, mesmerized by the tones of the blade, such high-quality metal as she had never seen before.

"Bro, it will be fun to rip these Angels apart with this."

CHAPTER 36

"It's so wonderful to hear your voice again, my bright boy. I have cleaned your apartment every single day."

"It's wonderful to hear your voice too, Nive. I'm very glad to know that my relatives did nothing to you."

"They just followed me home, and that was it. The following day I went back to my routine."

Antony was once again sitting alone on the big tire near the Dockbot, talking to Nivedita on his phone. It was twilight, and he looked tired after long hours working on the first robot.

In Antony's mind, Christine or Uncle Peter could hurt Nivedita in some way after the incidents in Brussels. However, after they made amends, he gathered courage to call her. Relief and happiness took over his heart and mind when he knew that everything was alright with his stepmother.

"I miss our conversations in the morning, Nive. These last days, I've been putting everything I have learned from you into practice."

"I miss you too, but I'm happy you left this depressive city for a while. Such a bright mind as yours deserves challenges. You can use your talents to help people, and please, forget the contempt you feel for your relatives," Nivedita's voice sounded serene and friendly.

"Yes, I have come to understand that in these last few days as well. Everything is so crazy. The challenges seem impossible to overcome, but I feel good... especially with Christine and Uncle Peter. We are getting along very well. Also, I've met great people here in this community. They are religious like you and Mom. You would love them."

"You seem happier, despite the difficulties. It sounds like a blessing. Angela and I have always told you that your uncle and cousins are hard people, and sometimes they seem not to have feelings, but... look at their lives... look at your life. You and your family are different. So, you should try to see them in a different light."

"I've just recently realized it. Well —"

"Every family has something good to offer, even if you disagree with a lot of what they have done."

"I don't know what I love more, you or your words. I have to hang up. Something amazing is happening here right now."

"Ok. Love you, bright boy."

"Love you, Nive."

Antony hung up as he watched Glenda leave the Dockbot, piloting a robot that had been completely rebuilt and plated with the new metal made from the metal taken from Muriel's armor. On its back, there was a long metal sheath that held a large sword. The robot's shoulders were equipped with missile launchers, and on its wrists, there were machine guns.

Many people from the community, mainly kids, approached to see the brand-new robot piloted by the best fighter. Everyone was amazed by its grandeur and how the metal, which was silver with soft lines of blue light, shone.

"Hey, if I were a contestant on a *miss robot,* I would absolutely win. Bro, these commands are soft like a pillow. You know very well how to arrange the gears. Be careful, kids."

Glenda took the big sword from the back of her robot and practiced some fast strikes. The weak sunlight that reflected on the metal and the sword blade created a spectacular show of movements that made the kids as excited as if they were in a circus.

“This machine is insane, Glenda. I’m glad to hear you say it’s easy to pilot.”

“Do you know how to pilot this toy?”

“A long time ago, when I worked in Cloud Base, I piloted some robots just to check the security systems. I just walked a few steps and felt like the best pilot in the world. But nothing compared to you. Well... I still don’t understand the blades on its feet.”

Glenda triggered two retractable blades on the robot’s right foot. One pointed from the heel, and the other one surrounded its right side.

“Remember bro, the *air kick*. With these blades on my foot this strike will be more than mortal.”

“Oh... I got it. That’s amazing. Well, guys, time to have a break. But remember that we have three days and two more robots to build.”

Antony talked confidently to the metallurgists who were around him with other people of the community like a true leader. These hard workers took a rest on some big pieces of junk and tires around the Dockbot while other members offered them some meals.

Glenda was still excitedly practicing some movements with her sword as the kids cheered her on, as if it were a real combat. Antony looked with glittering eyes, hoping that the end of the upcoming battle could be as happy as this scene.

CHAPTER 37

It was about 4:00 p.m. when soldiers and more soldiers camped around Cloud Base, surrounding a wide area, forming a huge military operation with tanks, cannons, army tents, and combat cars. In the sky, many drones of observation and equipped with guns flew over the camps ready for any anomaly that might appear.

On the top of Cloud Base, several missile launchers and some air traffic radars were checked by Christine Sims, who wore her habitual black blazer and trousers, accompanied by five men wearing military uniforms.

Each minute that passed saw more helicopters, jets, and trains arriving, bringing more soldiers, weapons, and military equipment. Every effort to protect the most sacred place on earth was welcome, despite the general manager of Cloud Base thinking it was an exaggeration.

Many contracts were revised so all the soldiers there would get good places in Cloud in case they died in combat — which Christine thought unlikely. *Two more days and we end all this trouble and go back to normal life,* that was what she had been thinking these last few days, even knowing she would watch the battle from inside her *precious* Cloud Base. *No matter what comes, we will defeat it,* she thought, and this confidence grew every time when she watched fragile Muriel lying down in the infirmary.

Muriel stayed in the same position on the bed, her wings a little more regenerated but with a sadder countenance after her last meeting with Antony.

On the top of the building, Christine continued walking with the army leaders, checking the new missile launchers, when suddenly a hard, cold wind blew west. Everyone felt the sudden decrease in temperature that made some strands of Christine's black hair lift up to the sky. She looked doubtful and nervous as she instinctively decided to walk to the edge, following the direction that her strands of hair were pointing in. Christine stood on the edge and noticed the sand of the desert being blown towards a dot up in the sky, about four miles from Cloud Base and five miles from the ground. This dot looked like a small live smoke ball which was attracting other clouds and some sand from the ground.

All soldiers around Cloud Base were impressed and scared, looking up to the sky, trying to understand something they had never seen before and which their minds could never have imagined. The rest of the soldiers camping in the back of the huge building left their positions to get a better view of the phenomenon that was appearing on the desert. Among train tracks, roads, army tents, and military vehicles, everyone looked hypnotized, looking up to the sky. Through the windows, employees inside the building jostled to get a good place to watch what was happening outside.

Many pieces of white clouds and sand joined up in the sky, making the smoke ball more alive and bigger. After some minutes, it stopped pulling sand up completely. Suddenly, silence took over the whole place, and this live dot up in the sky seemed to explode and become an immense hurricane of clouds, forming a terrible, huge Black Hole — as big as the Cloud Base building itself.

The day became night, and the cold increased more and more. The Army leaders gave orders to the troops and vehicles to stay positioned between Cloud Base and the Black Hole, which seemed to be a little inclined, as if it were ready to swallow the whole building. Gray and purple clouds appeared, flowing around, and bright strikes of lightning dazzled people as several peals of thunder made the scenery look even more apocalyptic.

Christine stood in the same place, frozen and astonished, unable to find a logical explanation in her mind for what she was seeing at that moment.

"I have two more days. Yes, two more days. To face what is coming," Christine whispered nervously.

The general manager of Cloud Base became brave, giving orders to the military leaders to position their troops for attack and calling for more efforts as she ran toward the infirmary. She crossed many sectors of Cloud Base, reminding all employees of the orders that nobody inside Cloud should know what was happening in real life. Since the incident with Muriel and Abaddon, all information that entered Cloud should be filtered by a new sector created in a rush, *The Filter Sector*. Many people outside loved to get in touch with their relatives that had passed away. However, the last disturbance had caused a panic, and it made Christine create a protocol to control information between the two worlds until the problems with the Angels were over. But this kind of control required a complex system, and most people who were living in Cloud continued without any information about the outside world. Besides, any visits and communications had been canceled for an undetermined period of time. *Maybe the world outside has been destroyed,* many people in Cloud thought. But most of the population there preferred to wait and keep enjoying their perfect lives.

Christine got into the infirmary with raging eyes and stood in front of Muriel, who still looked very weak.

The Black Hole appeared in its totality through the wide window.

"What will happen after this?"

"What?"

Muriel looked confused as Christine walked around the infirmary until she stood in front of the window, pointing at the Black Hole outside.

"After this? Did you know the size of this? It looks bigger than this building. How many of you are coming?"

"I don't know. At this stage, first the watcher comes. After that, my people organize the next stage."

"The watcher? What is that?"

"Look out... to the passage."

Christine turned and noticed that something big was slowly coming out of the Black Hole. For a moment, she thought her eyes were deceiving her, but no. It was a large white floating pyramid the size of a football stadium with enormous and viscous beige tentacles that, like those of an octopus, arose above it. On each side of it, there was a big eye with blue and green irises. This pyramid looked like it was made out of flesh and white stone. These eyes blinked slowly as it turned. It seemed to breathe like a ruminant animal ready to run, but suddenly it stopped and stayed static, floating on air. This pyramid stayed about two miles away from the ground in a space between the soldiers and the Black Hole, moving its long, thick tentacles randomly. Armed men, missiles, and cannons were pointed to the freak visitor. The men were waiting for orders from their higher officers, but nobody gave them; nobody knew how to act. They just stood there, nervous and trembling, expecting the worst, even though they knew that their good places in Cloud were guaranteed if they died.

It didn't matter that they knew where they would go after death. The fear of death was beyond their rationality.

CHAPTER 38

Inside the Dockbot, the work was intense on the second robot. This work was harder than the first one because the metalworkers were rebuilding an old machine, of which most of the parts needed to be changed. However, three people were frozen in the metal cabin, watching the weirdest scene they had ever seen on a screen. It was hard to believe what their eyes were seeing at that moment. Antony, Glenda, and William were astonished by the big pyramid floating on air as a crowd of soldiers stood below. Its big eyes reflected a distorted image of Cloud Base, and its tentacles moved like a monstrous octopus in the deep sea.

"Insane, more than insane," Antony didn't blink. "Did you talk to Muriel? Did she say anything about this... this... pyramid?"

Another small screen appeared in one of the wide screens. It was showing Christine's face, talking from Cloud Base.

"All she said was 'this is the Watcher, and he came here to prepare for the next stage.' Have you seen it in your dreams or something?"

"No, I haven't. Maybe the movement of armed forces around Cloud Base made them more careful about the next attack."

While Antony and his cousin talked, William and Glenda whispered in the back.

"Hey, Mom. Is she his cousin?"

"Yes, I guess."

"Hmmm... she looks hot."

"And she has good hearing, too," Antony added.

While Glenda held back laughter, William looked up, trying to disguise his last comment.

Antony turned off the transmission between him and Cloud Base, and his face showed bravery and doubt at the same time.

"Ok, guys. Things are becoming crazier than I imagined. So, we have only two days to prepare the best robots and choose the best pilots to fight. The time is coming for us to save billions of people and get our due respect. Also, William... come on, my cousin looks hot? The only way she'll look hot is if someone throws her in a campfire."

CHAPTER 39

It was 1:00 p.m. in a modern train station in Kansas City, where everything seemed to be going on as usual on a clear, beautiful day. Some trains arrived, and then androids and robotic forklifts unloaded and carried goods to other trains and trucks parked there. There were no humans involved in this process. All machines, including trucks and trains, were controlled by a simple system of Artificial Intelligence, and only a few employees, stationed at computers in cabins on top of the station, supervised every step of goods' transportation. These employees were the only humans that took any part in this process.

One of these employees, during his work routine, noticed that a red train pulling ten flatcars left the station without authorization — as if it had a life of its own. Many employees tried desperately to control this machine in different ways, but to no avail.

The train continued fast on its trajectory for one hour until it stopped in the middle of green fields. There, three giant robots were waiting for it. Glenda preferred to keep her robot with the original gleaming silver color of the metal, unlike Mike, who asked the metallurgists to paint his robot red, and William, who chose dark green, like the color of a soldier's uniform. Then each giant moved up to a flatcar and stayed there — each one separated by an empty flatcar. Some people from the community tied the robots with steel cables from their shoulders to hooks on the corners of the flatcars to maintain the balance of the tall machines.

The train started to move to toward Nebraska, leaving the people from the community behind. In the train, Antony, Glenda, William, and Mike were the only passengers going on a mission that could define the future of Cloud Base and, consequently, of Sims Corp.

Glenda and Antony sat in silence on top of the train as William and Mike Hong checked the cables on the last flatcars. Both guys talked happily, as if they weren't going on a suicide mission, and Antony just watched them, acting like son and father. Around them, an immense green landscape formed by fields and trees gave off a peaceful feeling that these people hadn't enjoyed for a long time. In this rare relaxed moment, Glenda realized that the gifted guy was miles away.

"Do you have kids?" Glenda took Antony by surprise.

"WHAT? ME? Oh, no... I probably don't have skills enough to... you know, to be a father."

"Well, once I saw some pictures on a gossip website where you were having fun in one of these playboys' parties with famous top models. So, I thought, humm... he probably has some babies around the world. Anyway, I saw that a long time ago."

"You don't look like the kind of person who stalks celebrities on the Internet, so... Ok, I slept with many girls in the past," Antony looked shy as he smiled, "but I got tired of that... you know, it's just an illusion. Maybe I am sterile too."

"Illusion? Come on, bro. Anyone would love to have your life. Parties, travels, no worries about tax and bills."

"So... imagine you have all kinds of pleasure in your hands every single day. But after that, what would remain? Believe me, all this can make you nauseous one day. In the end, only one thing could make you feel whole."

"And what could that be?"

"I don't know yet. My *mothers* say it could be religion... or God."

"Maybe. I noticed that you only talk about your *mothers*. Don't you have a father?"

"My biological mother, Angela, told me that I'm the result of a past relationship, and my father never cared to meet me. So, I have never cared to meet him either. She raised me to face it like a normal situation, but now... sometimes... I feel like I missed having a normal family and friends. Like yours."

"I just noticed you couldn't stop watching William and Mike. I admit I dated Mike after my Allan passed away. But only for a short time."

"Really?"

"Yes. Also, for me, you and William are very alike. Both of you are sarcastic and love jokes."

"If I were strong like William... instead of this chicken's physique that I have... I could believe that. To be honest... you know, I love to study ancient History... I started watching them because I was imagining the young guys who went to the ancient wars. The World Wars, Vietnam, Iraq, Eastern Europe. I doubt these guys were tense before the battles. It seems to me like they would just tell jokes between them because... you know... nobody can imagine what actually happens in a war. Because nobody is ready for it. Especially the young guys."

"I agree, bro. How do you train a teenager for something he can't even imagine? That's a good question. And one to which there probably isn't an answer."

"Right... anyway. Don't you regret choosing William for this mission? You know... something bad will probably happen."

"I have raised William to be the best and strongest. He insisted on going, ok. But I'd be stupid if I didn't believe in the person that I raised. Look at him, he is always so confident and smiling."

"And funny."

"Since my Allan and my other son Thomas passed away, I have always kept William with the more extroverted and funny guys in the community. Also, we made it a habit to always watch comedies, funny shows, and focus on our work. All this because I'd hate to have a son who had crying spells every time he heard words like *brother* or *dad*."

"That sounds wise... really wise."

"My Allan always said: 'If you can face your troubles with a smile, bro... you're indestructible.'"

Both stayed silent for a while, just watching the trees passing by fast in front of their eyes, waiting for the green landscape to become an arid desert.

William and Mike continued walking and talking happily on the flatcars.

The sky was blue, but Antony noticed something different on the faraway horizon ahead. A piece of the horizon was dark and full of lightning.

CHAPTER 40

Seven days after the fight between Muriel and Abaddon, a huge military camp was set around Cloud Base. All soldiers knew that the battle could start any minute. Nervousness, worries, and doubts took over the hearts and minds of the soldiers, whatever their rank or experience in the military.

Christine Sims stayed on the last floor, in a room where she could have a view of the whole battlefield. She watched every movement through binoculars and she was always in contact with the military officers on her phone. Also, this room was next to the infirmary where Muriel was still lying down, feeling sad and with a bad foreboding.

Suddenly, the Black Hole started making a noise like a kind of odd, loud roar, a mix of thunderstorm sounds and the sound of a whale. The lightnings became more intense, and the big pyramid kept floating in the same place. All these created an abominable scenery with a phantasmagoric orchestra in the background. The soldiers were trembling, holding their weapons, while others on cannons and in cabins inside Cloud Base, who were responsible for operating all drones and missiles launchers, kept their fingers shaking on their controls. *The last place in the world anyone could wish to stay* was what everyone was thinking at that moment.

In the cities and villages inside Cloud, people lived their lives like nothing dangerous was happening outside. Everyone had already forgotten the last incident and the phenomenon that had occurred in the sky.

The Black Hole continued its dark show of lights and horrific sounds until the soldiers noticed two big, winged creatures flying in circles, preparing to land. Both creatures flew toward the pyramid, then they landed heavily on the ground, causing an intense tremor and lifting a cloud of dirt around themselves. These creatures were about a hundred feet tall and stood for a while just watching the military camp — which looked like a camp of toy soldiers next to them.

Both creatures had the anatomy of a skinny human with their backs a little bent, long arms, and legs. However, despite their freakish posture, they looked brave and imposing thanks to their large shoulders, heavy armors, and cold weapons.

The first one wore a light brown armor that looked like it was made out of copper and a mask similar to the ones used by doctors in medieval times, with a long nozzle that made his head look like that of a mosquito. His large wings were made out of long gray feathers, and he was holding a long morning star. Besides this weapon, he had a long dagger with a silver handle on his waist.

The second one was a little smaller but looked more dangerous. He wore an armor with golden and silver parts, and an oval solid mask that gave him the appearance of a ninja. His long wings were formed by white feathers, and he held a long Arabic sword with a serrated blade. On his waist, he had a gray hammer with a golden handle.

There were just a few parts of their bodies that weren't covered by their armors. From the ground, some soldiers could notice that their monstrous bodies under the metal had a kind of lizard, dark beige skin with thick veins.

The two Angels stood frozen in a position of attack in front of the troops. Their wings moved slowly, closing on their backs until they looked like a compact backpack, which made the soldiers feel even more apprehensive. Christine, on the other hand, from the top of Cloud Base, was feeling more confident because she had been expecting more Angels.

"Just two. Big, but just two. I can finish them easily," Christine whispered. Then she gave the order in a severe voice: "Officers, give orders to the soldiers to open fire without mercy."

The high-rank officers replicated Christine's orders with only one word: "FIRE." It was the magic word that everyone had been waiting for to express their fury and tension through their weapons. All at once, hundreds of bullets, missiles, and bombs were launched at the Angels who remained static.

For a moment, everyone thought that the destruction of the targets was inevitable, but the two Angels acted in the most terrifying way possible. They leapt, twirling their bodies toward the soldiers with quick movements, dodging several shots. Some bombs and missiles hit them though they weren't enough to break their armors or stop them.

The coppery one used his morning star to hit the military vehicles, which were thrown through the air and exploded against the roads and train tracks. Each strike left a trail of destruction, fire, and blood. While soldiers were desperate, shooting their machine guns, this Angel, with just one strike, eliminated several trucks, cannons, and made many soldiers fly though the air — falling to the ground like bloody dead little birds.

The golden one was fast and ripped every military vehicle apart easily, like they were made out of butter. He spun his sword around without stopping, leaving on the ground many grooves filled with the soldiers' bloody flesh and pieces of metal from the cannons and trucks. Suddenly, he jumped and hit a massive column that supported a long bridge with a hard kick. The column fell, hitting other columns that, hit other ones, causing a *domino effect* — making the bridge fall over many soldiers below, spreading dirt, shrapnel... and terror.

"Missile launchers... FIRE," one of the high-rank officers ordered.

While a few survivors tried to escape the Angels' destruction or hid below the rubble created by the fallen train tracks, *missiles rained* on the military camp. Big sequential explosions happened in that area, making the surroundings of Cloud Base look like a forest formed by fiery pines and deafening noise. The giant warriors moved very fast with jumps, stunts, and sometimes hiding among the flames.

Some missiles hit The Watcher too, but it kept floating, stable, with only some flames shinning on its surface and tentacles — making its appearance even more bizarre than before.

Christine was astonished, just watching through her binoculars the sea of fire that had been created after the explosions of the missiles. No sign of the two Angels in this apocalyptic scenery. Only fire and wreckage.

"They died... oh, they died, I hope," Christine whispered.

However, the two giant warriors arose slowly from among the wreckage, moving their wings fast and spreading the flames around. The shots fired at them had only been enough to crack some parts of their armors, and the flames vanished little by little from their bodies with the movements of their wings. Then they closed their wings again and stood frozen like when they arrived, only watching the Cloud Base surrounded by piles of corpses, distorted metal, and the remains of bridges and roads covered in flames.

Only a few yards and a rail bridge supported by high concrete columns separated the giant Angels from Cloud Base's locked gates.

Christine put down her binoculars, desperate. She couldn't hold back the tears that soaked her cheeks and mouth. Her heart beat fast, and to endure her shortness of breath, she leaned on the glass of the window.

"If just one monster got in the Base so easily, imagine two giants like these," Christine whispered in a trembling voice.

The general manager was completely lost until she noticed, by her side, a train coming fast, carrying something that she could only identify as three more armored monsters.

CHAPTER 41

Glenda, William, Antony, and Mike stood on the train that was approaching Cloud Base. Desolation and desperation were the words that defined their faces as they looked at a scenery in flames, devastated by the two giant Angels.

"No, no, no... we are too late. Damn it," Antony was furious.

"Mother, look at those monsters. They're the size of our robots, and there are only two of them."

"And they haven't got into the building yet. We still have time to surprise them," Mike said.

"Mike is right. We still have the chance to save the day. I'm gonna stop the train in front of them. You have the power to finish this insane war. So... GO!"

Antony was still standing, controlling the train through his phone as the three pilots got in their machines.

The two Angels were walking slowly toward the train tracks — after crossing them, they would be in front of the Cloud Base's gates.

In the winged warriors' minds, that would probably be an easy mission, but they were surprised by a train that stopped on the bridge in front of them, carrying three metal giants. The Angels froze, just observing the robots ripping their steel cables from the flatcars.

"Ok, bros. Move your fat asses. We have two Angels to throw back to their damn world," Glenda said loudly.

The three robots leapt and landed forcefully on the ground, which was full of detritus and dust. They got into position to attack, then each pulled their own weapon out. Glenda held her long medieval sword, Mike pulled a sharp dagger out, and William wielded an ax with a long handle.

The presence of the robots made the two Angels look nervous for the first time, wielding their weapons as if they could smell the metal that was the same as that of their armors.

Tension took over the atmosphere among the fighters when Glenda explained her strategy:

"You two attack the coppery one while I go after the one holding a sword."

"Are you sure, Mother?"

"The coppery one is holding a morning star. A very heavy weapon. That means he's stronger and can take more punches. The other one is thinner, and if he's holding a sword, he's the faster one. And I'm the fastest of us."

"You're right," Mike agreed. "Let's go, guys!"

Angels and robots advanced at same time into an open and violent fight, showing no fear. A lot of dust and detritus flew through the air with each heavy step that the giants took.

William and Mike had no mercy on the coppery angel who had some difficulty protecting himself. While he defended against Mike's strikes, William hit him hard with his ax, mainly on his wings. Each of his strikes sent more gray feathers flying through the air. This giant was very strong, as Glenda had deduced, and he could withstand each strike bravely, keeping steady on the ground.

Glenda and the golden one exchanged sword strikes, making the clanking of their blades ring sharply. The Angel was fast and skillful, using his wings as a shield, though Glenda's sword ripped through them easily, spreading white feathers and blood through the air.

The metal replicated by Antony based on the piece of Muriel's armor was working and providing more resistance against the lethality of the Angels' weapons.

"They bleed... YES, they bleed," Antony said out loud to himself, following every movement of the two combats from the top of the train. Christine kept watching the combats as well, from a high room in Cloud Base.

The golden Angel pulled out his hammer, quickly striking at Glenda, who anticipated it and dodged the weapon. But the giant warrior didn't give up and advanced one more time, braver and faster, striking again furiously, with his sword and hammer at the same time. Glenda moved her legs fast and, with her great ability, managed to escape all the dangerous strikes. Suddenly, the Angel stopped his attack for a short moment, and Glenda took this opportunity to shoot the missiles from her shoulder, which exploded on her adversary. She took another opportunity as the Angel was stunned by the shots and hit him with a sequence of hard strikes of her sword. One of them hit the Angel directly on his mask, breaking it, showing one of his eyes. The golden Angel got away and stood in position for a new attack, as Glenda noticed his bizarre eye — a furious yellow eye, similar to that of a snake, surrounded by thick veins under dark beige skin.

"Bro... you are anything but an Angel."

They started exchanging hard sword strikes again, violently, in a fight that seemed like it would be defined by sheer force.

Antony noticed that down there on the ground, there were eight injured soldiers walking among the wreckage. So, from this moment, his instincts guided him more than his rationality. The thin guy ran to the next flatcar, threw one of the steel cables down, and went down as quickly as possible to try to help these men. The soldiers got startled when they saw the heir of Sims Corp in the middle of this disaster.

"Mr. Antony Sims, what are you doing here?" One of them asked.

"Forget the formalities. Let me help you. Here!"

Antony held one of them, almost unconscious and very injured, on his shoulder. Most of them walked with difficulty, carrying heavy machine guns. They slowly got away from the combat, and Antony tried to show the safer direction to follow, among big pieces of metal, wreckage from the columns, destroyed train tracks, and roads.

William and Mike kept striking the coppery Angel hard, and he seemed to show signs of getting weaker. However, he held his morning star high, spun his own body quickly, and struck down with his weapon, like golf player but more violent. It hit one of William's legs hard, making him lose his balance and fall. With the same strike, the Angel hit the waist of Mike's robot, who held the morning star firmly with one arm and stabbed the Angel's long neck with his dagger. William, with difficulty, got up quickly and hit the Angel's back with his ax.

"Time to leave this world, ugly face." Mike sounded angry as he whispered.

Through a brown mask that looked like a big mosquito's head, it was impossible to see the Angel's expression as he just bravely stood still.

The fight seemed to be over, but without Mike realizing, the winged warrior took the dagger from his waist and stabbed right into the cabin of the robot's pilot. The long and sharp blade ripped Mike's body from his belly to his neck, spreading lots of blood on the whole cabin, as if the Angel had stabbed the heart of the robot. Then the Angel held the dagger firmly in the middle of Mike's robot's chest, and with his other hand, he took one of the Robot's legs. William took his ax out of the Angel's back and held it up high, preparing another strike. But the Angel acted faster and stronger. He lifted Mike's Robot and threw it on William, who lost his balance and fell again. In that moment, William realized that

Mike's cabin was full of pieces of flesh and blood — which made him desperate and furious.

"NOOOOO… I'm gonna send you to hell, motherfucker!"

William tried to get up quickly, but his opponent was faster again. He held his morning star and struck William's robot several times in a row, mainly on the chest that became distorted with each hit. The Angel cornered Williams, who was propped on a pile of wreckage, trying to protect himself, but it seemed useless. Despite Mike's dagger still being stuck in his neck, the coppery Angel continued to fight vigorously.

Glenda kept exchanging strikes with her opponent, who seemed to be tireless. She had already realized that something very bad had happened beside her, but she also knew that one small distraction could cost her life. The pilot started feeling her muscles weaken and tremble. A strong tension took over her whole body in a way she had never felt before.

"This fucking monster seems unstoppable… and I can't find a weak spot," Glenda whispered.

Antony was on his way, taking the wounded soldiers to safety, but when he realized William had been cornered by a furious Angel, he made a desperate decision.

"Give me a gun. I must help my friend," Antony told one of the soldiers.

"Please, Mr. Sims. It's suicide."

"I don't care. Give me a gun now."

The soldier handed a machine a gun to Antony, who looked confused.

"OK, let me remember how to use this. I turn this on —"

"This way, Mr. Sims."

Another soldier, who held a machine gun, locked and unlocked his weapon, quickly showing Antony how to handle it. Despite Antony already having participated in sport hunting, this situation was making it hard for him to think straight.

"Ok, I already know how to use it. If you follow this path, you'll get to gate five. It's the safest way. GO!"

Right after he pointed the direction to the soldiers, Antony ran toward William as fast as he could. He was desperate and acting only on instinct. Also, he was so nervous that he didn't realize his heart was beating fast — faster than normal.

The injured and confused soldiers talked about the situation, trying to find an almost impossible solution.

"He's going to die. We should help him."

"Look at these monsters. We... we don't stand a chance against them."

"But I'll go anyway... I don't wanna feel responsible for a Sims' family member dying. Stay on the path, men."

"I'm going with you too."

Two armed soldiers ran back to Antony as the others kept on their way.

William was exhausted, enduring the hard strikes of his opponent with his robot's arms crossed in front of his cabin. After several attempts, the Angel destroyed one of these arms and hit the cabin so violently that it became completely deformed. Desperation and fear took over William's face when several distorted pieces of metal pressed on his body, crushing him, as he, tired of suffering, fainted, covered in blood and oil from the gears.

"Hey, guys. SHOOT HIM DOWN!" Antony ordered the soldiers.

The three armed men shot at the coppery Angel who felt nothing.

Glenda wanted to help her son but to do it, she had to decide her next movement quickly. After defending herself against a strike of his sword, Glenda kicked the golden Angel hard in his belly, and he fell back. It gave her enough time to run toward the Angel who was almost killing her son.

As the coppery Angel prepared his coup de grâce against William, Glenda jumped and spun her robot in the air like a metal ballet dancer holding a sword. The long blade ripped through the Angel's neck, throwing his head away along with the dagger that was stuck to it. The body of the Angel fell heavily on the ground.

Glenda hit the ground and tried very hard not to lose her balance. But any distraction could cost the victory. The golden Angel came fast behind her and ripped the arm with which she held the sword.

"HOOOO... fucking Angel from hell."

The Angel started another sequence of strikes which Glenda defended against with the only arm she had left, which was being destroyed.

"Come on... give me a break, so I can take a breath and end you."

One strike hit a missile launcher fixed on Glenda's robot's shoulder, causing a big explosion. Some flames reached the exposed eye of the Angel, scaring him and making him stand back a little. Glenda took this opportunity to kick his belly again, pushing him farther away so she could have time to take a breath. Despite that, her only arm became useless.

In that moment, the golden Angel was wielding a sword on one hand and a hammer on the other, and Glenda's robot had just one destroyed arm hanging from her shoulder by some steel cables and hydraulic plastic hoses. They moved slowly and carefully to their sides, just watching each other. Glenda was at a complete disadvantage, but the Angel looked like he had learned to recognize his opponent's bravery.

Antony and the two soldiers arrived under William's robot that was lying down on pile of wreckage. They got to the cabin and started separating and ripping out some pieces of the machine to try to take William out of it.

"Mr. Sims. This… this man looks like he's dead," one of the soldiers said nervously.

"NO, NO, NO... He's a strong guy. Look, he's still breathing. Help me get him out of here."

They struggled a lot to take William's body out of the cabin, completely soaked in blood and oil. Then, they carried him down the destroyed robot carefully.

While they carried William's body, Antony noticed that Glenda was cornered and without any advantage against her opponent. Then, he acted on instinct again, not realizing that his heart was still beating too fast.

"Ok, men. Take this guy as far away from here as you can."

Antony was walking back to the robot's cabin.

"What are you going to do?"

"Something crazy… Go, now!"

The robot's cabin looked almost completely destroyed, but Antony realized that the legs' controls were still working.

Glenda kept dodging the strikes from the golden Angel. It was very hard to find space or time for a good kick, especially with the piece of arm that just hampered her balance. Suddenly, she crouches down quickly and struck a sweep kick, but the Angel was faster again. He jumped and tried to strike Glenda with his sword. She dodged the sword but not the hammer that hit her violently between the cabin and the shoulder. Glenda fell on pile of wreckage unable to get up immediately and completely exhausted. Fear took over her mind when she saw the giant winged warrior with a furious eye approaching to give his coup de grâce. She breathed deeply and felt her life vanishing with each of the giant's steps she heard. Her time was up, and she was sure that her dramatic end had come.

"I can't stand this anymore... My Allan, I'm gonna see you... I hope... I hope so much."

The golden Angel lifted his sword, ready for the last strike. However, suddenly, Glenda heard the sound of other steps approaching at her side. William's robot, piloted by Antony, ran fast and awkwardly. His desperate shouts seemed to echo through the desert.

"GET AWAY FROM HER."

Antony threw the robot directly into the Angel, causing a violent shock, pushing the winged warrior away from Glenda. Both fell heavily, but only the golden Angel got up quickly as the robot piloted by Antony was squirming, fallen on the ground. Antony's heart kept beating fast, and he finally recognized the same pain that he had felt when he was playing soccer in the community.

"NO, NO. I can't pass out now. I —"

But Antony fainted and stayed unconscious inside the distorted cabin as the golden Angel approached. The strong-winged warrior forgot his other opponent for a while and prepared to strike his new opponent who was passed out. In the Angel's mind, Antony seemed more dangerous because he still had two intact arms, and Glenda looked weaker; he didn't realize the pilot of the robot was unconscious.

Glenda trembled when she saw the Angel lifting his sword to hit Antony. That's when, as if she was evoking a nonexistent power, the brave pilot recovered her confidence and decided to gather her breath to execute the ultimate strike. Glenda triggered the retractable blades around the right foot of her robot. Then, she got up and ran toward the Angel who was holding the long sword.

"That's my only fucking chance to finish it."

Glenda's robot jumped, and turning its body sideways, spinning in the air, threw her foot directly at the back of the Angel's neck. That was the *air kick*, the most difficult strike that a normal pilot could never accomplish. The blades on the metal foot ripped through the Angel's neck, and it gushed a lot of blood through the air, like a red geyser. The golden Angel froze and dropped his sword as Glenda fell to the ground. The winged warrior still took two steps back, then he fell over the robot piloted by Antony. The blood from his neck kept gushing intensely, covering Antony who looked like a red corpse inside a distorted metal hole.

"He died… oh, God… he died," Glenda whispered, relieved.

The victorious pilot watched her opponent, his eye still open but without life. Fatigue took over Glenda's whole body, covered in sweat and dust. She was unable to move for a while.

Christine Sims was sitting down on the floor still watching all the destruction through the window. She looked relieved as well, but breathing heavily, her eyes drenched in tears.

The pyramid with the tentacles started slowly moving back toward the Black Hole, floating until it vanished completely. This passage to the other world kept moving intensely, making gray clouds spin around in a completely dark sky.

In the infirmary, next to the room where Christine sat, Muriel finally got up and walked unsteadily toward the wide window, supporting herself on an IV stand. She hadn't recovered completely but could observe all the destruction outside with flames, corpses on the ground, and five giants knocked down. Even while watching this apocalyptic scenery, she had a hopeful countenance on her serene face and said:

"The humans are the victors... thanks to my Messiah. Now they have seven more days until the next... and final battle."

CHAPTER 42

Tokyo, thirty-five years prior.

It was about 1:00 a.m. when a luxury executive Airbus H255 helicopter flew among the skyscrapers of downtown Tokyo. The craft was white, and on its side, there was a blue infinity symbol.

Downtown Tokyo was lit by neon lights, and through its streets, fast sports cars raced at the same pace as their young drivers. The city pulsed with an intense energy, soaked in alcohol, and infested with all kinds of drugs. It was incredible how this city maintained its work routine during weekdays; when on weekend nights, it became an immense party orchestrated by electronic music.

The executive helicopter continued its trajectory above the bright streets until it reduced its speed at a bustling intersection. It landed softly, exactly in the middle of it, as drivers hit their brakes and curious, scared pedestrians stopped to see this craft, completely out of place.

The helicopter's door opened automatically, and a tall, burly man got out. The people around became more scared, including some cops who didn't have the courage to arrest him. This man was completely bald, and his deep black eyes contrasted with the albino skin on his face — that held a stern expression at all times. He wore a white suit, black shirt, and shoes — the look that marked his style. With firm steps, he walked, imposing, like a five-star general, along the street.

The crowd on the streets recognized him at once and was impressed with his young, rejuvenated appearance — because everyone around the world knew he was almost one hundred years old. It was thanks to the most modern techniques of rejuvenation and medical treatments that only he had access to and money to pay for.

This imposing man went inside a huge electronic music hall with a long line of people in front of it, but the astonished crowd just watched him go on his way. Not even the security guards did anything to stop him.

Inside the hall, a crowd of different ethnicities and styles danced as if there were no tomorrow. The dance floor was enormous, surrounded by many rooms and bars organized on three ample floors and illuminated by a powerful DJ light system. The whole structure was made out of metal and gray walls, though the lights were so intense that they created a psychedelic scenery where all possible colors in the world seemed to be there. In the back, there was a wide stage where two DJs animated the party from inside two big, bright cabins. However, when they realized who had just entered the place, they turned the music off for a moment. In that moment, only the lights and silence filled the place's atmosphere.

The tall man stopped for a moment, carefully watching the *colored crowd* around him, who looked apprehensive just watching him. After a few seconds of analyzing the environment, he asked in a deep and strong voice:

"Where is she?"

Some scared people close to him pointed up to the third floor.

He looked determined as he went up through an iron staircase until he got to the floor that had been pointed out to him.

In this spacious floor with no walls, there were several king-size beds, some of them separated by columns and glass partitions. Everything was metallic and silver, illuminated by neon lights inside and by the lights coming from the buildings outside. Boys, girls, men, and women, some of them almost naked, and others completely naked, had sex and did drugs.

On the first king-size bed of this floor lay, laid a beautiful, smiling blond guy, wearing only underpants, having fun with three girls. But his joy ended immediately when he noticed the big man approaching — he got so scared that he pushed all the girls away at once.

"Oh, you're here! I… I… I didn't bring her here. She came willingly."

"Where is she, Peter?"

"She's here, but I didn't bring her here… It's not my fault."

"Where exactly is she?" The man's voice sounded louder.

"Ok, ok… she's on the bed in the middle."

The man approached another king-size bed where a thin, small girl was sitting, snorting cocaine out of an old book called Trainspotting. She had long black hair with one long blue lock on the front right side. She was wearing thigh-high nine-inch heeled boots and a black and blue silk mini skirt. On her black top was stamped the cover of an album called Cross Purposes from an ancient band called Black Sabbath.

The girl noticed the shadow of a big person covering her, then she turned her nasty face up. She was wearing black lipstick and dark eye shadow.

Suddenly the man's voice sounded sad.

"Angela, what are you doing with yourself?"

"Nice to see you too, Allan," Angela's voice sometimes sounded sarcastic and sometimes angry.

"You could try calling me father sometimes."

"And you could try not to bullshit me anymore. Come on, asshole. Didn't you want me to make a baby for you to raise like your new toy? I have been trying these last few days. If you knew how many dicks have been inside me since yesterday... man... You'd feel so proud of me. Now, leave me alone... I'm busy right now."

Angela snorted another line of cocaine from the book.

"I just want you to come back home and... you know... live a normal life in a normal house and get a normal job."

"How about that blond motherfucker right over there? Do I have to obey you while my little brother is having fun around the world?"

"You're different, special. I know that, and you know it too."

"Yes... I'm different and special like everybody in our family. Everybody is a big, stinking piece of shit like you. A perfect fucking family that exists only inside your rotten mind."

"But you can try... try to bring this perfect family to reality."

"STOP, FAT MAN. I'm not another one of your fucking shareholders to be convinced. You've already destroyed your family a long time ago. Forget about me, please. Now, keep your focus on just destroying the world and let me be."

Angela snorted another line of cocaine and looked furiously at Allan, who looked lost and worried, standing there in silence, just thinking about what words to say next.

"I swear... I swear that everything will be different from now on. If you give one more chance."

"One more chance for what? Trying to destroy my uterus again? If you want a grandson to help you destroy the world, try making it by putting your dick in your own ass and —"

Allan punched the side of Angela's head hard, and she fell unconscious on the bed. He took his daughter and carried her under one arm, as if he were carrying a purse. Then he crossed the spacious floor toward the staircase, unaware that Peter was hiding behind a king-size bed, trembling.

Allan got to the first floor where everyone stayed in silence, just waiting for the unexpected visitor to leave the place.

Some people were recording and taking pictures of the most powerful man in the world with his unconscious daughter under his arm, as if she were a broken doll. Before leaving, Allan stopped for a while and looked at the people around him. In a loud voice, he said:

"In five minutes, I'm gonna check if these pictures and recordings are on the Internet. If I find any of them, I'll identify who posted them. After that, I'm going to destroy their lives... no mercy."

Some people at the party were deleting the pictures and recordings they had just taken. No traces of what happened that night were ever found, even on the deep web.

Allan calmly walked until he got to his helicopter, which had landed in the middle of the street. He got inside, and the door closed automatically. Then the craft took off, moving away from the skyscrapers as the sounds of electronic music from the party resumed with full power, as if none of that weird situation had ever happened.

CHAPTER 43

Antony was standing up on a large golden box, which floated on a sea of blood under a starless night sky. He looked confused, especially when he noticed a long crack under his feet and many desperate voices screaming from inside the box.

Suddenly, Muriel appeared up in the sky and landed in front of him. She looked brave, with stern eyes and dreadlocks down to her shoulders, resembling a lion ready to attack. She looked at her messiah as if she wanted to impose a terrible order.

Antony asked the Angel in a trembling voice:

"What... what is this place?"

Muriel just pulled out her sword and handed it to Antony, who held the heavy weapon awkwardly in his hands. Then she said:

"You know what you must do. So... break this box and let everybody free... free from their own sadness."

The screams inside the box became louder and more agonizing.

Antony was scared, looking at Muriel's brave face, who was just waiting for her Messiah to obey her order.

***

"What? Oh… It was a dream. Another bad dream… but… oh… Wait… what place is this?"

Antony woke up scared after another nightmare; this one completely different from the one he used to have since he was a child.

He was in a simple single bed covered in white sheets, in a little bedroom that was also completely white. There wasn't any furniture, carpet, or nightstands. Just a white bedroom with a locked window. Also, Antony was impressed with his physical condition and clothes. He was wearing his usual jeans, a tank top, his blue nylon jacket, and his sneakers were beside the bed. But what impressed him the most was that his clothes were totally clean, as if he had just bought them in a mall, and Antony looked like he had just taken a bath.

"I have never seen this place in my life," Antony whispered to himself. "It's not Cloud because, hooo… my head hurts a lot. But… this sound? This sound seems strange. It sounds like… a party! Or… a rave?"

Low sounds of electronic music came from all sides, and Antony tried really hard to figure out how he got to this place. He got up and put on his sneakers carefully. Then, still confused, he noticed that the room's door wasn't locked and decided to go out there. When he crossed the door, the sounds became louder, and the voices of a crowd could be heard as well.

Antony got to a long white corridor with gray floor, and there he walked warily, just following the sounds. There was a red double door at the end of corridor, which made him suspect what the place was.

"Oh, no… not this place… How did I get here? Well, I should go through these doors and find out," he whispered to himself.

Following only his instincts, but calmer in that moment, different from when he was in the battle in Cloud Base, Antony crossed the double door. Then, he was hit with an intense sunlight that hurt his eyes and the loud sounds of electronic music that made a deafening noise. When he recovered his sight, it was hard to believe what he was seeing: a wide camping site with pools, large tents, and a stage with a DJ cheering up a wild party with a crowd of excited people. Adults, teenagers, and a lot of pretty girls — everyone wore summer clothes, indulging in alcohol, drugs, and dancing wildly. Several people were taking pictures and short videos of themselves on their phones, while others danced inside the pools. A few yards away from there, Antony noticed a white mansion built with columns and pediments that made it look like a Greek temple. The moment Antony looked at the mansion, he knew immediately where he was. Also, at that moment, a smiling man appeared in front of him, holding a beer bottle.

"Hey, Tony, hahahah... Welcome to California. Have you had fun at the party? Come on, take a beer and enjoy the place."

"Hello, Uncle Peter... well I'm trying to drink less, ok?" Antony looked confused the whole time."

"Now is not the time to be shy. You saved our corporation, boy. Let's celebrate," Uncle Peter was extremely excited as the sound became increasingly louder.

"Ok, Uncle Peter... but... oh, this damn music... I wanna talk about Cloud Base," Antony said loudly. "You know, what happened and how did I get here? It would be better if we went to a quieter place. What do you think?"

"I can't hear you well... It's better we go to a quieter place. Come on."

Antony looked indignant and followed his uncle though a crazy crowd. While they walked, a group of friends, who were dancing together, crashed into uncle and nephew. Nothing serious, just normal trouble in the middle of excited people, but enough for the two to almost lose their balance.

"Hey, guys, take it easy, you're completely out of control hahaha," Uncle Peter said, smiling.

The Sims' family members got away from the area where people were clustered, leaving behind the noise, the smells of drugs and alcohol, and someone who was complaining loudly on and on, "where is my phone?"

Peter walked with one arm around his nephew's shoulders, always showing affection. Unlike Antony, who looked withdrawn, keeping his hands in his coat pockets the whole time. They were walking on a pavement that passed through some small grassy hills and continued toward the mansion. Antony noticed, on one of these hills, the blue Bombardier Global 1K.

"Hey, my nephew, remember this place? It's amazing, isn't it? Oh, before I forget. I left your *toy* over there. It's an amazing jet, boy. And look, I have flown in many business jets."

"Ok, thanks... did you bring me here?"

"Of course, I went to Cloud base to see your great feat. It was tragic, I confess. Destruction and carnage for all parts involved. Oh, terrible, terrible... But you won, we won. The people who died in battle are living happily in Cloud. We don't need to cry for the loved ones who lose their lives like in ancient wars anymore, you know, thanks to our beautiful Cloud."

"So... why did you bring me here?"

Uncle Peter took his hand off Antony's shoulder and threw his beer bottle far away.

"Because of this weird system inside your heart. I don't know very well how this thing works. It made you pass out and has kept you asleep until now. I don't know why. Then Christine and I agreed that I bring you here in your jet. Oh... don't worry about your pretty cats. They are in a box inside the jet just waiting for you. They are so cute. I have fed them every single day in there. You know, I don't like pets in my home."

"Every single day? How many days have passed since the battle?"

"Today it's been a total of... hmmm. Six days, I guess."

"SIX DAYS? Have I slept though all these days?"

"Yes, you have. I wouldn't want you to wake up at the Cloud Base infirmary. You know, all these robots, employees, and sounds of rebuilding. It's a bad environment for recovering. Ok, finally, we have arrived. I have something special to show you. You'll love it. Come on."

They went into Uncle Peter's luxury mansion, which had belonged to Allan Sims in the past. It was just a small part of an enormous property located in Orange County, California, formed by parks, pools, some wooded areas, and beaches.

While uncle and nephew got in the entrance hall of the mansion, good childhood memories went through Antony's mind. He remembered when he and his cousins Christine, Clifford, and other kids from family's friends played soccer all day long, and then everyone came to this hall to play hide-and-seek. Uncle Peter always came to participate in the game too, just for fun because he was always the first to be found, and the main reason for that was that he was already overweight — and because of that, he had some difficulty finding a place to hide from smart kids. In the end of these fun days, Uncle Peter would give ice cream to all the kids in the kitchen. These moments made Antony forget for a moment that he had been facing a bloodthirsty battle some days before.

After they went up and crossed some corridors, Antony and Uncle Peter entered a spacious luxury office with white walls and rustic wood furniture, including several bookshelves and two big rustic desks — one on each side of the room. The host sat down behind one of these desks with two wide screens side by side on a shelf behind it.

"Don't be shy, Tony. Sit down, please. We have a lot to talk about."

Antony looked suspicious when he approached the desk and sat down on a rustic old chair, keeping his hands in his coat pockets. On the desk, there were a simple tablet for making notes, a support for digital pens, a table lamp, a simple remote control, and a Headgate to visit Cloud.

Uncle Peter took the remote control and turned on the first screen. It showed a live feed from inside Cloud, displaying the street where Angela had just left the church where she usually attended.

"Look at my sister. She is an amazing woman. Do you agree? Of course, you do; she's your loving mother. To be honest, I hate this church she goes to. Once, she took me there, and I felt bad... oh, very bad... This ugly statue holding a baby and this book... a stone book on a stand, to me, it's all so freakish. You should kick the ass of the programmer who created this place."

"This is a statue of Mary holding baby Jesus."

"I don't know her. Is she one of our shareholders? She was probably important, I guess. Because to be honored with a statue... Anyway, look at my sister on the street."

Angela arrived at her house and sat down on the bench on the porch. Afterward, she remained in silence for a moment, just looking at the street. She took one of the books from a little table beside her and started reading. Just another normal day in Cloud.

Antony didn't like the course this conversation was taking, but he kept himself calm, just paying attention to his uncle speak.

"What's the point, Uncle Peter? Are we here just to watch my mother reading a book?"

"No, we aren't. The point is this."

Uncle Peter pointed the remote control to the screen and pressed a button. At the same time, Angela fainted on the bench and stayed lying down there, unconscious. Antony immediately got up, furious and desperate.

"WHAT THE HELL ARE YOU DOING?"

Uncle Peter quickly pulled out a pistol and pointed it firmly toward Antony, making him freeze nervously. His expression changed abruptly from that of an outgoing friend to a serious, cold killer. A strong wave of memories and doubts hit Antony's rationality hard in a second. *My own loving uncle, sometimes stupid, but still my loving uncle. How did it get to this point?* Antony thought.

"Take it easy, Tony, and sit down. It's more important than you imagine. And no worries... your loving mom is well, for now."

"For now?"

"Depends on whether you'll answer my questions correctly."

Antony sat down carefully, not even blinking, focused on his uncle, who seemed to be a completely different person from a few seconds before.

"Ok, Uncle. What do you want to know?"

"About Key. A damn system that you created to destroy our corporation. The *empire* that my father built."

"You have never cared about Grandfather."

"Don't change the subject, boy. I remember you threatened me and Christine. You told me that you hid a kind of button to turn off Cloud and make it just a big, useless building in the middle of the desert."

"Yes, I did it. After I realized the insane business between you, Christine, and Prime Minister Yelena. Your father created a place to avoid humanity's pain that death caused, not for you to choose who deserved to stay here or be caged in a computer program."

The cold stares of both seemed to warn of a tragic consequence.

"Welcome to the business world. Power attracts people who have power too. Thus, we have negotiated it. Now, show me the Key and destroy it or say goodbye to your mother."

"She's your sister! How could you?"

"Look at this place and all the legacy of my father. All of this wasn't paid for with familiar feelings; it was paid with money. Now, answer my question."

"If I were able to destroy Cloud, I would have already done it. That's why I got away from your damn business. You have negotiated lives instead of caring for them."

"Who are you to talk about philanthropy or charity? You have never helped others. You just isolated yourself in your luxurious apartment in Brussels and stayed there like a zombie. Also, let me show you another good video."

Uncle Peter kept his pistol pointed at Antony this whole time with one hand, while with the other hand he held the remote control. He pointed it to another screen that turned on and started showing a bizarre recording.

***

The recording showed Antony completely out of control, tied to a bed in an infirmary of Cloud Base. Uncle Peter, Christine, and some doctors were around him. Antony looked desperate, with his eyes red and dilated veins around his neck, trying to escape with all his strength from the ties, like a wild, enraged animal. His relatives and doctors watched him struggle like hunters who have no mercy seeing their prey die. Antony couldn't recognize himself in the video, and no matter how deeply he checked his own memory, he couldn't remember this moment.

Uncle Peter looked furious, asking Antony the same question, as Christine was confused, not knowing how to calm down her uncle and her cousin.

"Where is Key, Tony? I know you hid it, and I know you plan to destroy Cloud. *Yes*... say *yes,* damn boy. You hate your family. I know this... say *yes*."

"YES, YES. I HATE MY FAMILY. I HATE WITH ALL OF MY STRENGTH... I HATE ALL OF YOU."

"Uncle Peter, please," Christine intervened, "this is meant to be a treatment, not torture. This story about Key has gone too far. Look at him; Tony looks like a caged animal. Nothing he says makes any sense."

"Maybe, but there is some truth to his words. Look outside. The destruction, the smell of flesh and blood. He knows more than we can imagine. You tried to take him to Cloud, but you can't, because he locked the ship in his brain, and you told me that was impossible."

"Yes, he... he got it but I'm still working on it. Give me time."

"No, we don't have any more time. Do you remember what the winged woman said? Seven more days until a new battle... and —"

Antony became a little more relaxed but still trembled, speaking in a tired voice.

"We can't get into another battle... because we'll lose. We must free the souls... you must free the souls. The Angels won't stop fighting until they destroy this place. I... I was on a box, a box made of gold and inside it, there were people screaming. And an Angel told me that I should break the box to set all the souls free —"

"Look at him, Uncle Peter. You made him delirious."

"Wait, Chris, let me talk to him," Peter was calmer. "You say 'we can set the souls free.' Do you know how we can do it? Maybe you created something for that."

"Come on, Uncle," Christine whispered.

"Yes, yes. I created something. And this thing can destroy Cloud. This thing can set the souls free. This thing... can save everybody."

***

The recording was frozen on the unrecognizable Antony, tied to an infirmary bed.

Uncle Peter kept his pistol firmly pointed at his nephew, who looked nervous the whole time. However, from this moment on, he talked in a more confident voice.

"So, let me guess what happened. My friends and I won the battle, so you and Christine didn't need me anymore. Then, you two tried to... how could I say? *Shut me down*, take me to Cloud, like you and Yelena did to thousands of innocents in Europe. But since your agreements with Yelena, I locked the chip in my mind and got out of this damn business. I know it's dangerous to modify the chip, but I did it."

"Also, it would be easier to watch you inside Cloud. Ok, boy, show me where Key is, and I'll wake your mother up."

"I can't disable Cloud. You filled me with drugs and pushed me to say what you wanted. Now, do you wanna be right? Come on, Uncle."

"You said the truth! Like a drunk or a junkie who confesses his secrets... you confessed yours. Now tell me WHERE IS KEY?"

"All this time, the care you and Christine showed me was fake. You always loved to use people, to take advantage of them."

"You're the biggest fake here, Tony."

"For me... your gun is the biggest fake here."

With a sudden, quick movement, Antony got up from his chair, took the Headgate from the desk, and jumped to the middle of the office, doing a somersault on the ground, like a tire that came loose from a speeding car. Uncle Peter got up quickly too, and furious, shot at his nephew on the ground. This shot echoed terribly in the office.

Peter knew very well that to take Antony directly to Cloud, he had to hit a vulnerable part of his body, like the head or chest — but to do that, he had to have a good view of his target. So, this first shot, that hit the ground beside Antony, was just to scare him.

Antony hid behind another rustic desk to protect himself as his uncle stood, holding his pistol in that direction. Peter's face was red with rage, and the pistol was aimed at the desk the whole time, just waiting for Antony's next movement.

"Look at this fake gun, Antony. It isn't so fake now, yeh?

Peter shot one more time, and the bullet went through the desk, passing next to Antony, who kept crouching back there. The noise of the shot kept buzzing inside Antony's head and disturbing his thoughts.

"And don't you dare try another prank. I have an armed man just waiting for you right there in the corridor."

In the corridor outside the office, Joe, one of Peter's henchmen, stood with his ear at the door. He was holding a pistol, just waiting to get in according to the orders of his boss.

Antony's voice sounded desperate and entirely nervous from behind the desk.

"I know you don't wanna do it, Uncle Peter. You love me. You love your family... especially your sister. You are incapable of doing this."

Peter shot one more time at the desk, and again, another bullet went through it, almost hitting Antony. The noise was terrible, and the smell of gunpowder took over the air inside the place.

"INCAPABLE OF DOING IT? Hahahah... I'm incapable of letting my crazy nephew destroy the empire that my father built. You helped us make this corporation bigger, Tony. Now, why do you insist on making it so difficult?"

"It's already the biggest corporation in the world. Why make it bigger? Is it worth sacrificing your own sister for?"

Another shot went through the desk, making a big crack on it, and an uncontrollable fury took over Peter, making his voice sound diabolic.

"The last thing I ever thought about was your life and your mother's. Come on, Tony. You're disposable from now on, and your mother even more so. I'm giving you another chance to live with the rest of your family. Just end all this trouble and get rid of Key. Then I'll think about you two."

Suddenly, Antony got up abruptly with his arms held up, completely scared and willing to make amends with his uncle. On the other hand, Peter's face perspired hatred and lots of sweat as his neck's veins were dilated. He held the pistol pointed at Antony's chest with his two hands, his arms outstretched.

One foot of distance separated the two members of the same family, who could barely realize that their instincts could cause a tragedy in just one second.

"Ok, ok... I swear, I can't disable Cloud. I swear. You saw my face in that video. I was scared and so doped that I can't even remember myself in that state. I would confess anything."

"I know very well how Angela has raised you not to lie. You have said that you can't disable Cloud. But you never denied having created Key. These are your instincts respecting your mother."

"That's an insane reasoning."

"No, Tony. This is an insane reasoning."

Uncle Peter had no mercy as he shot Antony in the chest while he kept his arms up. Less than one foot separated them as the blazing gun barrel shot directly at Antony. A shot impossible to dodge.

CHAPTER 44

Inside Christine's office, silence predominated until the general manager of Cloud Base took an old telephone set and waited for an important call.

"Hello, Prime Minister? It's Christine speaking."

"Hi, Christine. Did you solve the problem about your rebel cousin? If you solved it, we don't need keep in touch through this prehistoric set anymore."

"My uncle insisted on taking care of it himself. Well, I don't know how it's going on right now, and, to be honest, my uncle isn't so intelligent. But I believe he's taking good care of my dear cousin, and this is probably our last conversation on this telephone."

"Very well, and congratulations on how you led the battle around Cloud Base. That *opened doors* for me to enter America."

"Thanks, Prime Minister. It was all under control. I was just waiting for my cousin to come and solve the problem. Very easy for me. Now we have every way to keep my... your plans."

"Of course. President Sowell has already granted an infinity of resources, and more than one million soldiers are ready to go to the Nebraska desert tomorrow. If the Angels destroy some of them, the American army will become weak, and I, along with other partners around me in Europe, plus Sims Corporation, of course, can control our planet's course."

"Won't the Prime Minister send more soldiers to America?"

"I promised it to the President. But I'm very good at creating pretexts to delay this process, like logistic problems or, humm... I can create an armed conflict here in Europe to avoid sending soldiers there... Anyway. President Sowell and his pathetic patriotic pride will try to solve this situation by himself. Poor man on his wheelchair. He's gonna kill his country, thinking he can save the world."

"If more big Angels like the last ones come here, that will probably happen. I have already protected the interior walls of Cloud Base very well, and Cloud Base Two is ready to work."

"Another Cloud Base? Where and why?"

"In the factory in Cheyenne, state of Wyoming, the same place where we are making more robots."

"Yes, I see."

"In the moment of the next battle, I'll isolate the Cloud Base's transmissions completely. Any big explosions outside could interfere inside Cloud, and hackers around world, including cyber terrorists, could take this opportunity to enter our system. You know, there are still people who wish to destroy or steal from my family by invading Cloud."

"Hmmm… good strategy."

"Cloud Base Two will work just to take the consciousness of people who get eliminated in the next battle. Nothing sophisticated. It's a simple backup until the battle ends. Then I can transfer all of them to Cloud Base. Also, all of the employees were transferred to Cloud Base Two and other offices in America… you know, *human work*, they are afraid, and I can't face a strike at that moment."

"To me, this last decision sounds weak, but I liked the idea of a new Cloud Base. It can be the beginning of new Cloud Bases in other continents. Europe, Asia, Africa. Have you thought about that?"

"Other continents," Christine sounded a little nervous. "Of course, I have. That's the plan, Prime Minister."

"Very well. We'll talk more about it after the last battle. When will you isolate the transmissions at the Base?"

"In a few hours."

"Are you sure your dear cousin can't enter the system?"

"Prime Minister, I guarantee that in a few hours Antony will be unable to interrupt anyone."

Christine Sims always tried disguise her admiration and envy of Yelena Orlyk. Her wish to become an equal or better than the Prime Minister crashed with her lack of talent to lead difficult situations like these. Cloud Base Two was just a temporary reservation, but Yelena just realized it was a good opportunity.

It didn't matter how much power someone has. Power was something addictive and hard to satiate.

CHAPTER 45

How many thoughts pass through our minds in one second? Hundreds of thoughts, but we are only conscious of one or two of them. The remaining ones are just possibilities created by us. It was a notion that went through Antony's mind as his own uncle was shooting at him.

Some years earlier, Antony imagined himself saving the planet and being admired by others, not because he was rich, but because he had acted right, because he was a good person. He could feed millions of starving people and distribute money to all of them, but he was sure this attitude wouldn't solve humanity's main problem. He knew very well that what should be changed was the way people understood the world and its challenges. Thus, the matter wasn't the allocation of food or money. The matter was how to improve the way people thought. So, humanity's main problem was inside their minds — the minds of the people who ruled the world and who just wanted a simple life. Antony had no idea how to solve this problem, so he decided to isolate himself in a depressing, monotonous life in Brussels. However, one day, freakish creatures he had only seen in his nightmares came to reality, making this rich, gifted guy push himself to act and look for a way to help the world. Antony interpreted all this weird situation as a call from the universe for him to finally do something good. But before trying to fix the world, he had to fix his life, and it included his decadent family — one of the main reasons that kept people from thinking differently and becoming better. So, all of a sudden, he had a strong wish to make amends with his family, because family was the foundation for any good society — and the foundation to fixing the whole world.

But if you realize your family is bad, will you discredit all families around the world? This was another idea that arose inside Antony's mind when he saw his own uncle shooting at him and the bullet just floating slowly in the air.

Uncle Peter was utterly confused and lost when he realized that the bullet he had just shot was floating between him and Antony — who looked very sad but serious and confident.

Antony lowered his arms and slapped the bullet that kept floating like a feather spinning in the air, which made his uncle even more astonished, staring at the pistol in his hand.

"What? What's happening with my gun?"

"Don't you recognize this place, Uncle?"

"This place? Do you mean my office?"

"No, Uncle Peter... We're inside Cloud. Actually, in a program that I created some years ago to interrogate terrorists and investigate criminals. I named this program *Mirror Effect.* A program that replicates in Cloud the place where a dangerous person stays in reality. Then, this program transfers him from there, without the bad person realizing it. Unfortunately, it was just a prototype because... you know... it's very dangerous to transfer someone directly from his mind's chip... without a Headgate. Also, it's impossible to take this person back to real life again."

"WHAT? YOU'RE A FUCKING LIAR. JOE, COME ON!"

"Your henchman can't hear you because his body outside is passed out while I lie down behind the desk with a Headgate on."

"No, it's impossible," Uncle Peter said, horror-struck.

Uncle Peter went furiously to the office's door and opened it determined to get out of this place. But when he crossed the door, he got in the office again where Antony was still standing in the same place. He crossed the same door several times and ended up back in the office every time, like magic. Peter couldn't bear his useless attempts to leave the office and turned back to Antony, shooting three more times at his nephew — and again, all bullets just stayed floating in the air like feathers.

"I almost forgot, Uncle Peter. This program can't replicate high-speed things."

Antony slapped all the bullets, making them float away. They just kept floating in the air like the first one.

"Damn boy. Let me out of here or —"

"Or what? You're gonna shoot me again?"

"How did you bring me here? I just want to know before I end your life."

"First, you can't end my life. Second... well, I love good explanations, so let me clarify the situation. Do you remember when we were crossing this crazy party out there, and some drunk people bumped into us? Well, I took the opportunity to steal a phone from one those guys and kept it in my pocket all this time."

"You had the courage to do it? So... you suspected me all along. No matter how I treated you."

"I really believed that our relationship was changing for the better, but instinctively, something still made me suspect you. I hoped my intuition was wrong, but unfortunately, it wasn't. So... while you were shooting at the desk, I unlocked the phone and accessed the *Mirror Effect* program in one of my files in my personal database. Anyway... then I transferred you into this copy of your office. Your adrenaline was so high that you were unable to notice this transfer."

"Damned boy!"

"The moment I got up from behind the desk with my hands up, we were already inside Cloud."

"Bravo, smart guy," Peter looked indignant. "Now what are your plans?"

Tears started to roll down Antony's face, who wasn't sure if he felt relieved to discover the truth about his family or hatred for having been deceived by them this whole time. He watched the image of his mother passed out on the screen as he talked.

"My plans were always to make my planet... I could say, *less bad* and love my family. Well, only the first one is left now."

Peter screamed with hatred and suddenly ran toward Antony and tried to punch him. However, he was so taken over by his own fury that he simply forgot where he was. His attempt to attack his nephew was frustrated because Peter went right through Antony, as if he was made out of light.

Antony looked very sad as he turned to face his uncle again.

"I hoped, one day, you would love your family like I tried. That was a stupid mistake. Goodbye, Uncle."

Like a light from a lamp that was just turned off, Antony disappeared, leaving an out-of-control Uncle Peter destroying every piece of furniture inside the virtual office and screaming nonstop.

***

Antony woke up and then got up, weak both physically and emotionally. He kept the Headgate and the stolen phone in his hands as he leaned on the destroyed rustic desk — taking some time to himself. He was still breathing heavily, unable to stop crying.

Suddenly, Joe, Uncle Peter's henchman who was still waiting in the corridor outside, broke the door and got in the office, holding a pistol. He became very nervous when he realized his boss was unconscious on the ground as Antony leaned on the desk crestfallen.

"FREEZE... hands up, man!"

Antony didn't pay any attention to the henchman and just pressed one of the phone's buttons on its screen, making Joe pass out at once. In that moment, the boss and his loyal employee were caged inside an office in Cloud.

Sadness and doubts still dominated Antony's mind when he remembered something very important.

"My mother! I have to go to my mother."

He put the Headgate back on and lay down on the ground, pressing lots of buttons on the phone in his hand.

A few minutes later, Antony was already in front of Angela's home in Cloud, watching his unconscious mother lying down on a bench. He approached her quickly and crouched by her side. Angela started waking up slowly and confused as her son caressed her head.

"Hey, Mom, it's me... Antony. Are you ok?"

"Hi, Antony. Yes... I'm... I'm ok, but I don't know why I felt so weak just now... and —"

"Yes, sounds good," Antony felt relieved, "but it's just a quick visit. I have some situations to solve out of here. See you soon, Mother."

Antony kissed his mother's forehead, who just closed her eyes. When she opened them again, her son wasn't there anymore.

Angela looked confused and worried at the same time. One week ago, some members of her family visited her every single day; however, these last days, no one appeared. Then her son showed up for a quick visit after she had mysteriously fainted. She stayed sitting down in silence on the bench, just watching some kids play on the streets with swords and shields made of cardboard. They were playing war.

CHAPTER 46

The crowd at the party in the back of Peter's mansion was still crazy, enjoying electronic music, alcohol, and drugs. They had probably completely forgotten their host, whose consciousness was caged inside a computer system while his body lay down in his office — and there he stayed for a long time.

On the small green hills of the mansion's park, Antony ran toward his dark blue Bombardier Global 1K, typing on the stolen phone. The jet's door opened as he threw the phone away and got in. In the jet's cabin, Antony noticed a cardboard box closed up with tape as the AI welcomed him.

"Oh no, that's insane," Antony was nervous as he took off the tape from around box.

"Welcome, young Mr. Sims. How can I help you today?"

"Go directly to Cloud Base in Nebraska, please, AI."

"Immediately, young Mr. Sims."

"I hope they're still alive," Antony whispered.

The Bombardier Global 1K took off, and in a few seconds, it gained enough altitude to speed up and fly fast, leaving the huge Sims Family's property behind.

In the jet, Antony opened the cardboard box quickly, and inside, he found the three moggie cats looking so tired and weak that they couldn't even meow.

"'I have fed them every single day'... Uncle Peter wouldn't stop lying."

Antony looked for something to eat, for both himself and the cats, inside cabinets and the fridge. There were several kinds of cereal bars, sandwiches, bottles of juice, milk, water, and alcohol, all preserved by a refrigerator system powered by the nuclear battery of the jet. Antony ground some cereal bars into a bowl, and the cats started eating them with the last of their strength. While Antony fiercely devoured some sandwiches and drank alcohol at the same time. He filled a bowl with milk for the cats, who that little by little looked happier and started to meow slowly.

"Hey, my friends, don't worry. It'll all be ok from now on," Antony said as he petted one of them.

Six days after the battle around Cloud Base, Antony was kept in a coma through several kinds of medicines and drugs, and at that moment, he knew nothing about what was happening with Christine, Glenda, and Muriel. Inside Cloud, everything seemed to be the same according to what Antony saw when he quickly visited his mother. But that was it. He was still lost about his uncle's actions. *Imagine what Christine is preparing for me,* it was the horrible idea that arouse in his mind. However, that was not his main concern, but Glenda and William were.

Antony found a laptop in one of the cabin's compartments. In all vehicles and properties of the family, there were modern phones, tablets, and laptops, especially in places where Antony attended. He sat down in an armchair and started checking what could have happened during his absence these last days.

The blue jet flew over Los Angeles downtown, which in the past had been an example of modernity, entertainment, and good vibes. But over the last decades, several poorly elaborated immigration laws facilitated the entrance of people from Latin America, who always thought that their cheap labor would guarantee their lives there. They were more or less right, but all hotels, business buildings, and restaurants became accommodations for poor immigrants and transformed the modern downtown into a big slum — guaranteed by the same bad laws. Unlike New York, in Los Angeles, wealthy people lived in their secured properties around the hills of the city as immigrants took over downtown. Despite the laws permitting the entrance of immigrants, there were no laws or any other form of assistance for these folks, and their numbers just increased, filling the streets with dirty small shops, food stalls, brothels, and drug dealers. At night, not even the police had the courage to transit there, a forgotten place only created solely chaos, strengthening bad ideologies, violence, drug trafficking, and making even richer the people who secretly invested in these kinds of businesses. Actually, concealed agreements between political leaders and powerful corporations had as a goal to transform many cities into these kinds of areas where thugs and bandits reigned. *The plans of WPU worked perfectly there, creating chaos in cities and then offering poor social assistance that only tricked the folk into an illusion of better days. That's the easiest way to control nations around the world. But one day all of that will finish. Everyone down there, including the crazy folk partying in Uncle Peter's mansion, is going to die, and their souls will go to a big USB drive called Cloud... no one else will be born, and Angels will destroy what will remain of the Earth. I'm in the middle of a silent extinction process.* Antony thought, feeling sad, watching what once had been a beautiful city, a reflection of what the whole world had become — a world he wanted to save.

After a moment of silence watching the city below, Antony turned his attention back to his laptop. Sims Corp had most of the satellites spread around the world, and Antony could get access to every one of them to get a panoramic view of any part of the world he wanted to see. The first place he looked for was Cloud Base, and that was his first surprise too. The Base looked normal, like nothing had happened these last few days.

"It's impossible. Rebuilding all that area that was completely destroyed in just a few days?" Antony whispered. "Also, where are the missile launches from the top of the building? These images aren't in real time."

Between biting a sandwich and a beer gulp, Antony typed on his laptop quickly, accessing other satellites and getting the same images of Cloud Base. That's how he realized that someone, probably Christine, had applied a program called *Mirage* to the area. This program worked by fooling the satellites, covering the real area with an old panoramic picture of it — like it patched up a hole on a map with the old picture of the damaged part. This kind of program was used to cover up devastated forests, huge polar ice caps melting, and areas occupied by armed forces. All situations where politicians and powerful businessmen needed to negotiate territories and hide military operations, *Mirage* was used. But the most impressive part was that Antony couldn't break this program and see the reality under the fake image.

"Well… I'm either too old to break this program, or Christine has done her *homework*. I can't access it. Let me check the community."

The second surprise was to realize that the folk of the community had vanished. All houses and Dockbots were empty and there weren't any robots around.

"No, no... What is happening? At least three thousand people lived there. They can't simply vanish. I'm gonna check the days past."

Antony could access the recordings of the satellites too. Through them, it was possible to check the movement of people from some days before. He noticed an intense traffic of vehicles and employees of Cloud Base, carrying equipment from the Dockbots and leading the folk to buses. All of that happened one day after the battle.

"These airplanes... hmmm... let me check where they are from."

Through the satellites, Antony could scan the vehicles and find their origin and destination.

"Cheyenne's factory in Wyoming. Hmmm... interesting. What was happening there?"

After some commands on his laptop, Antony accessed images from this factory, and again the *Mirage* was being used on this place. And again, Antony was unable to break the program. The images just showed a calm factory — they were probably recordings of weeks past. Also, Antony tried to call Glenda, but her phone was nonexistent according to his search.

He looked confused, though more relaxed for at least being in control of his own situation at that moment.

"Damn... Christine got to hide everything from me. She was probably sure that Uncle Peter would fail. What do you think, my friends?" Antony looked at the cats.

One of them replied with a meow.

"Hey, you're looking better. Would you like to visit a new place?"

Another cat replied with a meow too.

"Nice. Ok, AI, change of plans. Now, we have to go to the Sims' factory in Cheyenne, Wyoming."

"All right, young Mr. Sims. Changing the course right now."

CHAPTER 47

On top of Christine's large desk, the general manager checked, watched, and controlled everything in Cloud Base through a touchscreen system. At that moment, she observed the arrival of new giant robots made in Cheyenne — imposing machines similar to those Antony had built with Glenda in the Dockbots of the community, but more modern, standardized, and armed. They were all white, and the only difference from one robot to the other were the red numbers printed on their shoulders, like a military rank patch. These robots arrived at Cloud Base by trains and heavy lift cargo helicopters.

Another huge military camp was being organized, bigger than the first one... much bigger. About two thousand robots, military vehicles including tanks, missile carriers, and cannons were positioned for an upcoming battle. Hundreds of military tents of different sizes and functions formed a kind makeshift city in the area between Cloud Base and the Black Hole — a combination of human soldiers and robot soldiers controlled by artificial intelligence or by other soldiers stationed in a cabin many miles away from there. Unfortunately, these kinds of soldiers didn't have the same efficiency and reasoning as humans, leaving the hard work to the *true soldiers*.

It was a starless night when the huge Black Hole appeared up in the sky, and there it stayed for a few days, causing fear and tension in thousands of soldiers who were camping around Cloud Base. Drones flew near the Black Hole, and some of them tried to fly across it, but all they found was the same black sky, as if it were transparent. Only the creatures from the other world could use this kind of gateway; the humans had no idea what could exist on the other side of the Black Hole.

Christine Sims was focused on her work when suddenly, a red alert blinked on one of the several screens on her desk. She looked a little startled as she tapped this screen and dragged it to the middle of the desk. With her fingers, she zoomed in on this screen a little and realized something that made her angry and nervous at the same time. This screen showed a live feed from the infirmary where Muriel was hospitalized.

With another tap on the desk, Christine called security:

"Security, send guards to infirmary number five, quickly. I'm going there right now as well."

Christine left her office and crossed the corridors until she got to an elevator, which would take her to the floor where the infirmary was located. Her face was tense, like she was losing control after some days managing to keep her plans working well. The elevator reached the floor where the guards were waiting for her. Christine walked fast with firm steps to the infirmary, walking before the guards. They understood they should follow her just by looking at her angry face.

When they got to the infirmary, they noticed the wide window had been shattered, and the bed was empty.

"Damned monster. She will probably come back to attack us. But I'll be ready for her," Christine whispered.

The general manager of Cloud Base and her guards stayed in silence, just watching the winged woman fly toward an immense Black Hole.

***

Muriel, completely recovered, with her armor and sword intact, was flying through a dark sky and being watched by many kinds of drones. She dodged each one of them, flying elegantly and invigorated by her gray wings that moved slowly. Some drones suddenly changed their trajectory and started to chase her, probably following orders from Christine and the armed forces commanders.

The Angel didn't pay attention to the many drones that were after her when she entered the Black Hole, and inside it, she disappeared — leaving the drones to fly through the anomaly up in the sky to find just the same dark sky on the other side. Miles from there, in some cabins of a HQ, the drones' pilots were indignant for not being able to find out what was on the other side of this mysterious gateway that only Angels could access.

On the other side of the Black Hole, Muriel flew through a gray sky, spattered with several long and thin clouds, like waves on a vast gray ocean. Below her, there were many dark mountain peaks that arose from an immense thick fog. This environment looked like an infinite sea of mists with black static icebergs — a depressive landscape that showed both peace and death at the same time.

On one of these peaks, Muriel noticed someone standing, wearing a long black robe and a hood that covered his face, like a grim reaper. In his hand, he held a scythe with a long sharp blade, and at his back, he had two silver wings, long enough for their tips to almost touch the stone ground. Around his waist, there was an old thick rope which adjusted his robe and held a long copper hunting horn.

Muriel looked confused when she caught sight of this familiar being just watching the Black Hole on the horizon. She flew down in a spiral and landed softly behind this dark Angel who didn't seem to worry about her presence.

"Cassiel... are you still waiting for me?"

He turned to Muriel and took off his hood. Cassiel had the appearance of an old man, no hair or beard, just an albino face that only showed peace, and mysterious red eyes with black catlike pupils. Around his head were two gray thick horns, swirly like those of an old antelope. His voice sounded grave and calm, like a priest teaching something important in a confessional.

"My lovely Muriel... I would wait for you through eternity. You know that."

Muriel got very emotional and walked directly toward him. She hugged him tightly, like a daughter who, for a long time, hadn't seen her father.

"You did not need to wait for me. You should be with the others, preparing for the next attack."

"I am too old for wars and the noise of warriors training."

They let go of each other slowly after a long hug.

"Now I am just the one who will announce the beginning of the next two battles," Cassiel said.

"Two battles? My Messiah can still stop the last battle. We should wait more... like our God taught us," Muriel looked worried for the first time.

"My lovely Muriel...You should admit your Messiah can fail. You saw him. He changed his mind, and he is not on our side, not anymore. I saw it through the eyes of The Watcher. Your Messiah should lead the humans to destroy what our God named The Cemetery of Souls, the name which you put inside the mind of your Messiah, but he is leading the humans to protect this terrible place."

"No... he changed his mind, but he can come back to our side. My Messiah can help us again. I saw the goodness within him, and I feel it deeply within me too."

"Feelings that do not result in actions are empty and useless. You cannot enter the dreams of your Messiah since our worlds are connected by the gate. Therefore, you cannot see within him clearly. Our time is merciless... like our God," Cassiel sounded like a loving and severe father.

"Our God wants our worlds to live connected in harmony. Because of that, I led my Messiah up to here."

"Your Messiah did nothing except for making an inevitable war even more devastating. He cultivated hope in humanity; now they are stronger... But you know the destruction of the place where the souls are caged... The Cemetery of Souls... is inevitable."

"Yes, I know... my Messiah can set free all *Lives* that are caged in there. For a long time, he has hated this place, and he is the only one on Earth who can destroy it."

"I have so much hope in that... unfortunately, most people on Earth suffer through the decisions of the few they call leaders. If your Messiah does not stop the first war, the second will destroy the whole planet."

"You mean an *Apocalypse*?"

"Yes... An end to those who don't learn the truth and a new beginning... a new age to try learning again."

"But if we just destroy The Cemetery of Souls and set free the souls from there. We don't need an apocalypse. Apocalypses are terrible for the universe."

"Terrible but necessary for the equilibrium of the universe. The problem is not just The Cemetery of Souls. The problem is the idea of it.

If we just destroy this place, the humans will build more... and more. The actual Cemetery of Souls is not a place made by humans but an idea in their minds, a horrible idea to control life and death. If they don't destroy The Cemetery of Souls by themselves or if we cannot destroy this idea in their minds, we must destroy them... to destroy this idea."

"Yes... yes, I know... I completely forgot about that. All that has been happening these last few days is puzzling me, I have never felt this way before," Muriel looked down, holding in her anger.

Cassiel approached her and carefully lifted her face with his hand.

"My lovely Muriel... You are doing what the goodness within you is pushing you to do. That is why our God chose you for this important mission."

Tears fell from Muriel's eyes and rolled onto Cassiel's fingers. She looked sad, unable to express her feelings with words, when suddenly she noticed someone flying over them. A man who had long silver wings and a muscular body with light purple skin landed softly near the two. Muriel got emotional again when she saw the one who was so special to her and ran over to him.

"My Abaddon."

"My Muriel... I missed you so much."

Both Angels hugged warmly like a passionate couple who had been apart for years. Then they let go of each other but kept holding hands. Muriel caressed Abaddon's wide chest and looked at his face through teary eyes.

"I thought I killed you... I am so sorry."

"You were following the wishes of our God... There is no reason to say sorry. And you thought a sword to the heart could kill me," Abaddon smiled, "my love for you is stronger than that."

"My Abaddon... You always amaze me."

"My warriors," Cassiel spoke up, "the time is coming, and we must be ready for the next battle. Abaddon, show Muriel our new plans."

"New plans... What do you mean, my master?"

"Do not worry, my Muriel... follow me and I will show you."

Abaddon stepped away and took off with a strong impulse of his wings that spread dust around them. Muriel was a little scared when she looked at Cassiel, who just nodded seriously at her. Then she took off too and followed her partner until she reached him up in the sky.

The Angels flew side by side over the mountain peaks, almost entirely covered by a thick fog, against a cold breeze that came from an empty horizon. Only silence and loneliness prevailed on that landscape. Muriel and Abaddon looked like two small dots crossing an immense gray and white ocean where life didn't exist.

"What did Cassiel mean when he said new plans?"

"You will see, my Muriel... Your Messiah is an incredible human, but I think he is not acting like you wished... Because of that, we had to change our plans."

The fog dissipated little by little, and the mountains became more visible when both Angels decreased their speed and landed on the edge of a hill. There they stayed, sitting down on a big rock, just watching a terrific scene down in the valley. The whole landscape was gray, made out of stones, rocks, dried up rivers, and a few points where there was a sparse vegetation.

Muriel looked sadder and hopeless as she watched what was happening down there.

"I do not believe our God has allowed this... It seems like all our people are here."

"This is necessary to keep the balance of the universe, my Muriel. People from Earth are killing our people by keeping their own *Lives* caged on Earth... and it is affecting other worlds after ours. If your Messiah doesn't act quickly... an apocalypse will be necessary."

Down in the valley, thousands of different kinds of Angels were getting ready for war, like an immense camp full of medieval warriors. There were hundreds of giant Angels like the two who had attacked Cloud Base and thousands of Angels like Muriel and Abaddon, every one of them sharpening their swords, spears, and daggers or training hand-to-hand combat with each other. The variation of color of skins and shapes of the Angels' armors were infinite — white, green, red, beige, and red skins covered in armors which varied from gold to silver, iron, copper, and several other kinds of metal. Faces covered with masks shaped like dragons, jackals, tigers, skulls, kabuki, and goats — simple but deformed helmets and other bizarre ones made this crowd the most diverse army Muriel had ever seen.

On the faraway horizon, behind that crowd, there was a long and high fence made out of thick tree trunks with lit torches on top of them. Behind this fence, the scene was even more impressive. Another army of angels, ten times bigger than the first one, was also getting ready for war. The landscape was covered by bizarre and dangerous warriors as far as Muriel's eyes could see. A countless number of warriors who disappeared on the horizon.

Floating over this place, an enormous pyramid with many tentacles and with four eyes on its sides kept spinning slowly. The Watcher looked like it was doing nothing, but it could analyze through the universes millions of possibilities and situations and share its ideas directly to the consciousness of every Angel. Since its first visit to Earth,

The Watcher saw the humans' courage against the Angels and the danger that represented to the universe's balance.

Muriel was astonished watching her people prepare for a battle that could destroy the two worlds.

"Look at that, my Abaddon! Our whole world seems to be here."

"The kids and thousands of women are safe on the other side. When Earth is destroyed, they will repopulate the two worlds."

"You talk as if everything were already over."

"You should admit it, my Muriel... your Messiah only encourages the humans to fight, and it will bring a tragic end to Earth."

"No... he is a good one. He just got lost in his mind for a while. I have followed him since he was a child, and I believe in his goodness."

"But what if you are wrong? The humans killed two of our best warriors... everybody here is afraid because of it. And I know what you saw when you left The Cemetery of Souls."

"The humans name this place Cloud Base... and yes, I saw it... I saw many metal giants around it. Giants created by my Messiah."

"Now, you understand why we changed our plans... We hoped your Messiah would destroy this... Cloud Base, but he is protecting it, and he allowed the humans to kill two of us as well."

Tears rolled down Muriel's eyes as she looked at the crowd of Angels getting ready for war — men, women, teenagers, and the elderly too, all together and focused on training sword fighting, defense with shields, and spins in the air with their wings open. The giants trained with less intensity, wielding swords and spears with thick spearheads.

"What is the meaning of the fence in the distance? The crowd beyond it is immense."

"On this side of the fence are the warriors who will attack first... beyond the fence are the ones will attack later."

"But the first group of warriors seems enough to destroy the humans around Cloud Base. Why a second group? And why so many?"

"The plans of the humans remain unchanged. Did you not see? You know, I am fast, and I hide very well. There is not just one group of human warriors... There is another one beyond the warriors around The Cemetery of Souls, coming from the northeast of Earth. A bigger one... much bigger."

"That is impossible. Can they sacrifice themselves for this place? A place used to cage *Lives*?" Muriel looked disgusted.

"Yes, they can... I saw in the northeast an immense group of human warriors and their weapons. Metal carriages that spit fire, big metal eagles that launch *explosive spears,* and warriors... thousands of warriors and their weapons that spit fire very quickly. A crowd as big as ours."

"And what is the plan now?"

"Our first crowd will attack The Cemetery of Souls. It will be a carnage, I know, my Muriel... but, even with a lot of difficulty, we will be victorious."

"And then?"

"Then... the second crowd of human warriors will arrive at The Cemetery of Souls. When it happens, our second crowd will cross the Black Hole and attack against them furiously. This way —"

"An Apocalypse arises... Our God cannot allow it... And if my Messiah stops the first battle?"

"I hope your Messiah does it... I hope so much. But it seems he will do nothing to prevent the war," Abaddon looked worried seeing his love's sadness.

"I was not mistaken with my Messiah. I... I believed in him so much. I dedicated my whole life to this mission. I cannot believe that our God would do this to me."

Abaddon caressed Muriel's face, moving her dreadlocks out of the way a little and wiping some tears from her eyes that kept staring at the crowd below in the valley.

"Maybe our God wanted you to make a mistake. We almost killed each other in His name... now we are here. If our death had been the wish of our God, and we knew that before our fight... we would respect Him the same way."

"Yes, I understand, but for me, the humans do not need this terrible end."

"Remember what Cassiel told you... the problem with humans is most of them suffer because of the few that lead them."

"I remember... I realized that when I was on Earth. Few of them rule their world. But these few ones are bad or became bad because of the excess of power, I am not certain... that is the terrible sickness of Earth. The good ones who want to lead and fight for a better world do nothing and spend their lives just complaining about the bad ones who have power."

Muriel leaned her head on Abaddon's shoulder and stayed this way, crying, her eyes fixed on the crowd getting ready for the upcoming war. Abaddon leaned his head on Muriel's head and caressed her hands, which held onto him tight. The silence between the two Angels in love prevailed, and just the sounds of metal from the Angels training down in the valley echoed through the landscape. Both Angels looked sad because they were sure that it didn't matter if they had good intentions; they couldn't stop the worst end for the two worlds.

CHAPTER 48

It was a sunny afternoon, the sky a spotless blue, but the blue color of the Bombardier Global 1K was darker, and it was easy to catch sight of it flying over the vast green area in Cheyenne, the state of Wyoming, where one of the Sims factories was located. The Sims factories were spread around the world and were responsible for providing anything Cloud Base needed to keep working — computers, reactors, generators, and all kinds of vehicles like cars, trains, and airplanes. They produced, in partnership with other businesses, everything that everyone around world used in their homes and at work. However, there were some exclusive factories of the Sims Family over which only the main members had total control, like Christine in Cheyenne.

Antony looked a little apprehensive, sitting down in an armchair, just watching the green landscape below through the window. One of the moggie cats was lying down on his legs, and the other ones were on the ground playing with each other. They were all healthy after the days they had spent confined inside a cardboard box.

Suddenly, something odd outside approached the aircraft. Antony noticed something flying in the opposite direction of the jet but not too far away. There were about twenty heavy-lift cargo helicopters carrying white, giant robots, heading toward the Nebraska desert. He became indignant watching the robots, which were very similar to those he had built with Glenda in the community, though all of these ones were white and more modern and armed.

"Hey, I created this kind of robot. Damn it, it's insane. They are probably going to Cloud Base. Which means that there is another battle coming up," Antony whispered.

"Two minutes until we arrive at your factory in Cheyenne, young Mr. Sims," the jet's AI notified.

"Thanks, but I don't have a factory. Only a crazy family that wants to take over the world."

In two minutes, Antony saw an area full of standardized square white buildings in the middle of a vast green pine forest. Far away, the work site looked like a toy city made out of big white cubes organized in lines by an unimaginative child. An intense traffic of trucks, heavy-lift cargo helicopters, androids in small vehicles, and trains going toward Nebraska made this huge place much busier than it would have been on a normal day.

The Bombardier Global 1K landed on one of the heliports around the buildings as a small car, similar to airport baggage carts, approached. Antony looked suspicious when he got off the jet and saw the car, driven by a beige and gray android. This kind of android was called Jack, like the ancient expression *Jack of all trades*. They had humanlike anatomy, made out of iron like the Tin Man from the *Wonderful Wizard of Oz*, but thinner, and on its oval head, there were just a camera lens and a small speaker through which he spoke in a polite male voice. The android Jack model was the cheapest and easiest android to program for all kinds of hard work. That's why it was the best seller around the world, and, of course, Sims Corp dominated the manufacturing and trade of these machines.

"Good afternoon, Mr. Sims. How are you? It's probably all amazing. Please get in the car so that I can take you to meet General Glenda Jackson. She's already waiting for you," the android's voice sounded very happy all the time.

"General Glenda? That doesn't sound good. Not good at all... Well, ok," Antony looked suspicious constantly.

Antony got in the car, and then the android drove through a winding path until reaching an ample area where there was an intense traffic trucks, small cars, and forklifts. More lift cargo helicopters were leaving the factory, carrying more giant robots. Also, he noticed a long conveyor belt coming from one of the buildings which slowly brought more robots to be taken to Cloud Base.

All this situation in Cheyenne was possible because Christine used the time that Antony was in a coma to steal all the technology he had created some days before to fight against the Angels. Then, the general manager of Cloud Base used all available technology to make hundreds of robots on a large scale. This was what Antony concluded as he watched all this commotion of heavy vehicles, but he had no idea how Glenda had become a general. Maybe she already had this military rank before they met each other, despite Antony not noticing anything characteristic of someone who had been in the army, and Glenda never having mentioned anything about a time in the military. However, after Uncle Peter tried to kill him, Antony didn't trust anyone. She could have been hired by Christine or Uncle Peter or even Yelena to spy on Antony and use him to build this new defense for Cloud Base, even though it seemed so unlikely to him. Nevertheless, of one thing Antony was sure: the last battle was over, and there was still a terrible war coming up. If not, there wouldn't be any reason for all this work in Cheyenne.

Suddenly, the robot driver started a conversation with Antony during the trajectory.

"It's an amazing sunny afternoon. Do you agree, Mr. Sims?"

"Yes, I see... Where does Glenda... I mean, General Glenda stay?"

"In the last building just ahead, Mr. Sims. After working hard the last few days, training the new robot pilots for the war, General Glenda Jackson took today to organize the war strategy. You know, Mr. Sims, the war is coming, and the victory of the Sims Corp too."

"Training pilots?"

"Yes, actually training soldiers to pilot these amazing robots to fight against the winged monsters who are coming for war."

"Winged monsters! Christine still calls them that... hmmm. Ok, but does she know of my arrival?"

"Miss Christine Sims has been monitoring your jet. After she noticed your arrival, she notified General Glenda Jackson immediately, as previously ordered by the general."

"Glenda giving orders to Christine? That sounds odd," Antony whispered.

"We have arrived, Mr. Sims."

The android stopped the car in front of one of the white buildings.

"General Glenda is waiting for you on the seventh floor, in room seven. You may take the elevator in —"

"I know, I know. No problem," Antony left the car hastily without paying attention to the robot driver, who just watched the young man walk through the wide automatic double doors of the building.

"Have a lovely afternoon, Mr. Sims. When your visit is over, just let me know."

Antony walked firmly through a wide and empty entrance hall and got in a big elevator. Inside it, he said *seventh floor*, then, automatically, the elevator went up.

He was very nervous and determined to find answer for the questions that buzzed inside his mind. *What if Glenda is a traitor? And William? What happened to him? Maybe he is inside Cloud*. Antony couldn't hold all these questions in for much longer.

When the elevator got to the seventh floor, its doors opened, and Antony could see an ample meeting room with a long table surrounded by armchairs. Some laptops and papers were spread on the table, and on the wall next to a wide screen was a map of the center of the U.S.A with several red dots on the Nebraska desert. The office was almost empty, and the lights were off. Only the sunlight illuminated the room.

At a wide window, at the far end of the office, a tall woman wearing a camouflage army uniform and cap looked at the vehicles and robots moving outside. Antony recognized Glenda, who had her back turned to him, and he started feeling a mix of worry and fear of being so close to discovering hard truths. He approached the supposed general and stood behind her. She noticed his presence but kept her eyes on the scenery outside. Also, he noticed that she no longer wore any makeup or earrings.

"Glenda."

"Hey, bro... you look well," Glenda's voice sounded very serious the whole time.

"You too. I didn't... I didn't know you were in the military."

"I was never in the military. The stupid people who run this damn place made me wear this."

"Why?"

"They thought a uniform and a rank could help me get the respect of other soldiers as I trained them to fight using the robots."

"Oh... nice... well, you won the battle, we won the battle. You are important to everybody and —"

Glenda finally faced Antony. She looked angry and indignant as she interrupted him.

"And what? The battle! The Angels! That's all you worry about. Did you even think of asking about my son? Do you remember him?"

"Of course, I... I remember, I am very worried about him... I was going to ask if —"

"He's in a coma, and I don't know if he'll go to a true heaven or your damn Cloud. My old friend Mark is in this damn Cloud, but... Can you guess? I can't see him. I can't see my best friend. Also, I can't see my son because a monster almost smashed him, and his case is so grave that he can't leave the ICU. And what did your damn cousin propose to me... I mean, forced me into? 'You save my damn Cloud Base, and I'll save your son's life.' All because of your fucking battle."

"Hey, take it easy. You always knew the risks for you and your son. We had long conversations about that. Now you can't complain. I always worried about you," Antony was nervous, but he wasn't intimidated.

Glenda took a phone out of her uniform pocket and started typing on it quickly.

"Always worried about us? Hmmm... So, can you explain this?"

An audio recording of Antony's angry voice rang out from Glenda's phone, which both could hear:

"The media is saying nothing about it; there are just people going to Cloud, but there haven't been any births in the last ten years, just fake edited news with old recordings of babies, creating a damn illusion that there are babies somewhere on this planet. It's clear to me... the world population has decreased about one billion while the population inside

Cloud just grows every single day... and you're fucking worried about that? You just talk about that like it was a kind of planet reaction. Of course, Cloud is to blame for that, but... damn... —"

The recording ended abruptly. This was part of a conversation between Antony and Yelena Orlyk during their meeting on the bullet train over the Nemo River, minutes before Antony saw the Black Hole for the first time.

The revelation about the decrease in population had been a hard blow to Glenda, especially since she had spent hours talking to him about her life, and Antony never mentioned this horrible detail about Cloud. Also, she still remembered the moment when she didn't understand the reason why, for such a long time, nobody had been born in the community. After this recording, Glenda understood that the problem wasn't just in her community but worldwide.

Antony knew it had been a terrible mistake to hide this information from his new friend, and that it would be hard to apologize or change Glenda's mind, especially after he knew William was on the verge of dying. Moreover, Antony couldn't have imagined that his venting in front of his enemy could affect a friendship that didn't exist yet at the time.

Glenda's face trembled with repressed fury as Antony's seemed frozen.

"You lied to me," Glenda continued, "I still remember that you looked awkward when I talked about the lack of kids in my community, and I thought, 'he's an odd boy, that's all it is,' but no... you were hiding something from me. Something you were sure that if I knew, I would never agree to help you."

"It's... it's not the whole conversation. Can't you see it? Yelena is... you know, manipulating you, that's her plan. Turn us against one another," Antony was desperate.

"And she did it because it was fucking easy, because you know this fucking computer, building or... whatever that your grandfather created is keeping kids from being born. I would never agree to defend this damn place, but now they have my son as a hostage... and now his life depends entirely on them."

Tears rolled down Glenda's face as she turned back to the window, just to avoid facing Antony, who had no words to comfort his friend. His teary eyes looked outside too, as if the sky could bring some relief to his repressed desperation, but no. Antony always found, very quickly, solutions to any problems that could arise suddenly. But in that moment, he was feeling powerless, because for the first time, he wasn't facing a problem about technology, but a problem about a true friendship. Antony remembered that in desperate moments like this, he always acted by instinct, and he never found a good solution.

"I can... I can change it. I can save the world, I swear —"

"Stop, bro... STOP... you can change nothing," Glenda spoke with rage, "give up while you still have time, before you hurt more people... and yourself too. Unfortunately, the world is in the hands of bad people. You have to admit it and you get away from it."

Antony gesticulated frantically and hysterically with wide eyes. He was losing his self-control increasingly with each sentence that he spoke.

"And what can I do? Let the world end while I enjoy my life on a paradisiac beach surrounded by girls and alcohol?"

"If I were you, I would do just that. Get away from this shitty life, bro," Glenda looked calmer. "You can still save yourself. If part of the world ends, you will be in the good part with girls and alcohol. Let people like me suffer... as always."

"No, no... I won't act like the people I have always hated. Leaving the world to explode and people to burn while they live in the security of their castles, no, no... I can change it, we can change it. Everything will be good like... like —"

"Like before? Life was never good, bro. And you are one of the guys hidden in castles just smiling, watching poor people burn... gosh, this conversation makes no sense. Please, little chicken, move your fat ass and get out of here."

"Yes... I'll go, but I'm going to change all of that and save you and William too."

Antony left the place with firm steps. In a few minutes, he took the elevator and got into another car driven by another Jack. He was visibly out of control, nervous, with one of his hands constantly on his chest, expecting a possible fainting spell, and trying to calm down with deep breaths. His only goal at that moment was to arrive at Cloud Base and solve everything with his cousin; then, he could face Yelena and President Sowell. However, for now, he didn't have any plan or idea how to save his friends and the people inside Cloud, including his loving mother.

Glenda stayed close to the window, just staring at Antony in the car, going through the parking lots of the factory towards his jet. With tears still rolling down her face, she watched the blue jet fly toward an uncertain future as she heard an incisive voice from behind her.

"Did you dismiss him like I told you to?"

"Yes, I did."

More and more tears fell from Glenda's eyes.

CHAPTER 49

The blue Bombardier Global 1K vanished into a deep horizon, and Glenda couldn't take her eyes off the spot the aircraft had been only a few seconds earlier. The movement of the machines outside, an upcoming war, and the moving from her community to a different place, chosen by Sims Corp, had changed Glenda's life completely in one week since an odd guy had come into the place where she lived and asked for help. She would never say no to someone asking her to save the world, but in that moment, she was sure it had been the worst choice she had ever made in her life. The most powerful corporation in the world took everyone she loved as hostages, and if she didn't follow their orders, she couldn't even imagine what might happen to them. They gave her a military uniform and a fake military rank she hated, but this corporation wasn't the only problem she had to face.

While Glenda stayed at the window, just looking outside, a woman, taller than her, came in and stood beside her, looking serenely out the window as well.

"Yelena!"

"Prime Minister Yelena or Yelena Orlyk to you, General Jackson."

"Just Glenda, please. I never trained or studied to be in the military."

"But you have the skills for it, I admit."

Glenda maintained her expression very serious and continued to look through the window so she could avoid facing the Prime Minister. Unlike Yelena, who talked in a calm and imposing voice, as if she had total control of the situation — which she really did.

"I would like to say 'nice to meet you, bro.' But we have already talked a lot by videoconference, and to be honest… it's bad to meet you. I thought you would watch the war from your comfortable home in Europe."

"I really admire women who speak honestly. Anyway, the events that are coming make up the greatest enterprise of my whole life, and I can't just manage it through videos or calls. I have to be here to start this new era."

"Do you think we can win?"

"No, I don't."

"So, why care about this... like you said, this enterprise?"

"Winning or not, I'll get some advantages. With the USA completely destroyed, my Europe will rule the world again. I know you understand a little about ancient History. After the First and Second World Wars, my continent was completely destroyed, and this country gained a lot from this destruction, gathering immense wealth. This enterprise will be a good opportunity to change the game."

"Oh gosh, it sounds like damned revenge. So, me and the other soldiers, we are only cannon fodder as you bring your evil plan to a close like a movie villain. Have you considered that these monsters might destroy Cloud Base?"

"Of course, if it happens, I'm planning on building more. To be honest, it's already happening."

"Building more? I thought only Antony's family could build it."

"Yes, only they can. But after this war, I don't know if they will be as powerful as they are now. Thus, I'll enter their business. They can't monopolize this technology forever."

"A monopoly that you would love to have in your hands."

"Of course, this war will bring me a lot of opportunities, as all wars always do. I'm really excited about it."

"I can't believe that. What about the babies? Was it true, what Antony said? No one can have kids anymore because of this fucking Cloud?"

"In my opinion, that is just a coincidence. After Cloud's population reached three billion, women around the world seemed to become sterile. Fake and crazy theories have been scattered around by the scientific community. A new virus, pandemic, anomaly in human evolution. Anyway, the theories are infinite, but there's not one exact explanation. Because of that, and to avoid causing panic in the whole world, governments and big corporations, through the media, of course, have striven to hide this... this little hindrance."

"You call human extinction a *little hindrance*? Gosh... You don't have kids?"

"I don't have time to lose with such insignificant things that will be solved soon. For now, I let the media do its good work."

"Good work? You mean lying to people."

"I like to call this *a temporary way to blind the population*. You don't know what the media guys are capable of. They edit old interviews, documentaries, and TV shows where several kids appear, then broadcast it as if they were current."

"And of course, you're loving it," Glenda looked disgusted.

"I admire the media's inventiveness to create so many different ways to lie to people. Most people think the lack of kids is a local situation but... you know very well... it's a global situation. At least there are people like me who see this little hindrance as an opportunity."

Glenda finally stared at Yelena who stayed calm and sometimes let a sarcastic smile escape her lips. Unlike Glenda, who looked indignant. They faced each other, as if at any moment they could start a fight.

"What do you mean by opportunity?"

"Hmmm. You look curious. I love it."

"I was just keeping to myself here when you arrived and started talking nonstop... gushing out your superiority. Now, spit it out!"

"Well, your skinny friend blames Cloud for the sterility of women around the world, but he forgot to mention that there are several kinds of artificial insemination that can replace the natural ways of reproduction. Imagine the possibilities, we can create stronger, more intelligent humans who are immune to more diseases. In short, a new race for a new era. I know you're a religious person. I can say it in a way you'll understand better: God gave up on humans, but I can... create them again."

"If you can create humans, you can choose their positions too."

"Can you explain better, General?"

"You understood me very well. If you can control all births, you can control what these new humans will be like, their ethnicity, their lives. Can you have an example? Of course, you can. Kids from ethnicity *A* will be born to rich families and will be intelligent, while kids from ethnicity *B* will be hard workers without capacity to study or understand how the shitty world they live in works. Am I right?"

"Hmmm... you might be right. You're smart despite your origins."

After a few seconds staring the Prime Minister down, Glenda took a deep breath and continued.

"Where does this thirst for power come from?"

"Since a caveman grabbed a large branch and held it up in front of several other cavemen. He imposed his power over the others with this gesture, and even though they were in a larger number, the others respected him."

"You don't look so old to remember that moment."

"Don't be sarcastic. You understand me very well. People tend to admire people who show power and follow them even without being sure whether they are right or not."

"That sounds so stupid."

"It's not. Just look at history. Power isn't something that people conquer. It's something freely available in nature. Whoever realizes that first, gets it. If they don't, someone else will come and get it. And another one and another one... So, I'm just holding up my branch and I won't drop it so soon."

"It's unbelievable. Gosh!"

Glenda felt completely hopeless as she turned and looked at the horizon through the window again.

"Do you still miss your skinny friend?"

"He saved my life and the life of my son too. Gosh, this little chicken saved my life. Have you told him what he really is?"

"No, I haven't. It's a family matter. It's Christine's business."

"It'll be all over for him. He certainly won't get over that."

"He is what he was born to be. It's up to him to choose how to face it. And yes, he probably won't be able to get over it. Anyway, I'm loving our conversation, but we should go back to work. Send me the reports from the last training with the robots as soon as possible. Then, be ready to go to Nebraska. I'll be in my office upstairs."

Yelena left the office and, as usual, she just gave orders without saying please or thanks.

Glenda stayed at the window a little longer, just watching more and more heavy lift cargo helicopters carry giant robots. She was sad, but she couldn't decide if this feeling was about herself or her friend who was flying away.

CHAPTER 50

The Sparta Military Base in the state of Wisconsin, previously named the Sparta Military Reservation, had for a long time in the distant past been Fort McCoy. Fifteen years before, this place had been totally rebuilt, and a huge green area had become nine hundred thousand acres of pure cement and asphalt, like an infinite shopping mall parking lot. That's why this military base had been given a name similar to its former name, Sparta, not as a tribute to the ancient Greek city-state and its brave warriors, but because past governments had decided to replace names that referenced heroes of American society. Names like Abraham Lincoln and Benjamin Franklin were less known than the current artists of showbiz, only remembered by a few people in the country. Also, Names like Robert Bruce McCoy, the national guard honored by the name of the fort, had been completely forgotten. Therefore, it didn't matter what the actual name of this place was since it wasn't a tribute to any important people of America.

Thousands of soldiers, women, men, and Jacks, the androids programed for military service, prepared tanks, trucks, and jets to go to the Nebraska desert and face the most bizarre war they could ever imagine. In the last few days, the human soldiers had seen many videos and speeches about what they would face. Employees and soldiers of Cloud Base had shown them pictures and footage of the battles between Muriel and Abaddon and the giant Angels versus giant robots. At first sight, they were terrified to the point of giving up and leaving their positions in armed forces; however, the members of Cloud Base and especially President Sowell calmed them down, explaining that this time the army was better equipped with missiles more powerful, and bullets made out of the same metal as the Angels' armors. Of course, the discovery about the metal in Angels' armors had been made by Antony in the Dockbot in Glenda's community, but this information was hidden by Christine.

Every soldier looked crestfallen, each one accomplishing their responsibilities with competence. They knew very well that they were the last frontier between the security of the nation and its destruction, even though that could be their last mission in the name of their country.

Along this huge place, among hangars, quarters, and garages, there were some big digital screens, similar to those from movie theaters, with loudspeakers fixed under them.

Everybody kept working hard, sometimes talking and joking with the androids, though without losing focus on their responsibilities. Some of them took a rest in vehicles or on top of containers, but the anxiety seemed to have such an effect on them that they would wake up quickly to go back to their work, acting automatically just like the Jacks. Sometimes, the rhythm was so frenetic that it was hard to distinguish men from machines.

The afternoon sky was cloudy, and a cold breeze blew around the base, making the vibe even more depressing, when all big screens turned on, and the face of the President of the United States of America, Richard Sowell, appeared. He was in the Oval Office at the White House, sitting down on his usual wheelchair at the ancient and well-preserved Resolute desk, a gift from Queen Victoria of the United Kingdom in the XIX century. His gray suit, red tie, and round glasses made the President a character easy to recognize by every American. The President's green eyes showed bravery and serenity in that moment, like a warrior ready to attack but still aware of his body's limitations. Despite that, his voice sounded paternal, like a wise and calm father sure he was giving an important life lesson to his kids. The President's speech rang through the loudspeakers and was listened to carefully by everybody, including the Jacks in the Sparta Military Base, who stopped what they were doing for a while to listen:

"Dear Americans, we're going to face the worst and strangest war we have ever gone through. To be honest, to me, all wars happen this way. This war didn't begin because supernatural creatures decided to invade our world. No. This war began here by forces we already know very well. Forces that have wanted to destroy our concepts of family, religion, and freedom. These forces know very well that these are the three pillars that support our society, and if they destroy these concepts, they can easily control a lost and alienated society. These evil forces' actions brought these supernatural creatures here to try and fix the world in the worst way possible, and everyone here has paid for the egocentrism of a few... unfortunately. So, dear Americans, all of that has taught us that everything you do for your kids, partners, your parents, your friends, and our neighbors is valuable and can affect our whole world. This war was brought here because in the past, we didn't know how to take care of each other, because we didn't know how to choose good leaders who knew how to lead our society bravely. Now, we shall fight, and we shall take care of each other in the battle and after the battle. Because we'll win, and we'll go back to our homes to be better... better fathers, better mothers, better sons, and daughters... better friends. Because if we must fight in a war to become better people, we shall face it; we shall destroy all kinds of enemies who come from the sky or from the earth too. Like a wise former leader said in a difficult past: we shall fight on the seas and the oceans, we shall fight with growing confidence and growing strength in the air, we shall fight on the beaches, we shall fight on the landings, we shall fight on the hills, we shall never surrender. So... dear Americans, please... keep this in your hearts and minds: God will always be on the side of those who will win."

The transmission ended, and all the screens turned off. Everyone stood for a while, looking at the sky with a special spark in their eyes that made the cold breeze turn into a gust of hope. Little by little, they went back to their work, but more motivated, even the robots.

One hour later, this place was completely empty.

CHAPTER 51

His fingers moved fast on the laptop, trying to find some way to enter Cloud Base's system, though Antony was always coming back to the beginning, which made him more nervous and out of control. He and the three moggie cats were inside the Bombardier Global 1K, flying fast, leaving the state of Wyoming behind and crossing part of Colorado, toward the Nebraska desert.

As he tried to enter the system, Antony whispered his doubts:

"If I get to enter, what am I going to do? There is a big battle approaching, and I can't close the Black Hole through my computer; and if I stop the robots, I can destroy Cloud Base and kill everybody there, including all the soldiers and Glenda."

Sometimes, Antony seemed to talk to himself, and sometimes to the three cats, which were relaxing, eating some cookies and pieces of bread on the floor. The whole time, he sat in an armchair just looking at the screen of his laptop with his eyes wide.

Antony knew very well that he needed to talk to Muriel, but after their last meeting in Cloud Base's infirmary, he didn't have the courage to ask for her help. He thought she would help him stop the war, but this Angel had come over just to confirm that inevitable end.

There is a solution, there is a solution, he tirelessly pushed his mind on with this affirmation.

"If my mother were here, she would have an answer... No... Damn it, Antony. You're not a child anymore, you're an adult. You can solve everything, you can solve everything," Antony whispered.

Antony kept typing when suddenly a video call opened on his laptop's screen. It was Christine, talking from her office in Cloud Base. She looked serious but showed sarcasm and confidence in her eyes. Also, she had an unusual smile on her face, unlike Antony, who was completely awkward but trying with all his strength to look serious and controlled.

"Hey, Antony."

"Christine? Hello... I've been trying to contact you. It's good to see you."

"No, Antony. You weren't trying to call me. You were trying to enter Cloud Base's system without my authorization."

"Your authorization? I never needed you to enter any system."

"Right now, you do. Finally, I can stop you from entering. It took me a while, I confess. But I got it, and you'd better follow my orders at least once in your life."

"You don't remember I learned to use a computer by myself when we were kids despite you being older than me. I don't believe you did it."

"You always believe you are the smartest of the family. So smart that you left us for nothing... to be drunk every single day. If it wasn't for me and Uncle Peter, you would have never kept your comfortable penthouse apartment in Brussels."

Christine wasn't the loving cousin she had been days ago, before the battle in Cloud Base, anymore. She got angrier and angrier, always reminding Antony of his mistakes and how irresponsible he had been with the family business. Also, she didn't mention, and never would, that she'd needed the help of about twenty high level computer engineers to create the kind of cyber barriers that would keep Antony from invading the Cloud Base's system again.

"You always needed me, every single day. You, Uncle Peter, Clifford, all of you could never do anything without me. That's why I deserve a penthouse apartment in Brussels. Because of your damn inability to solve problems by yourselves. You always need me to keep this corporation up. Admit it!" Antony drooled as he spoke.

"I feel so sorry for you, Antony," Christine looked indignant but calm, "if only you had followed us, followed your family, you could be a normal rich young man enjoying a good lifestyle, but no. You preferred to think that you were the most intelligent person in the world and act differently from what you had been programmed to act like since you were born."

"That sounds so stupid. Programmed for what? I'm not a computer system."

"You're worse than a computer system. Did Angela never tell you? Of course, she didn't."

"Leave my mother out of this!" Antony looked increasingly out of control.

"No, I will not because she is the main responsible for this. Have you never asked yourself why Angela is the only religious person in the family? Well, who would suspect a loving religious mother? Maybe that's what religion was created for. To control people and hide their evil intentions behind an angelical facade."

"Don't bullshit me —"

"Or why Clifford called you *abortion leftover* when we were kids, or why you've never seen pictures of your mother when she was young? Maybe it's the same reason why you have never found your father? I know Angela never told you, and you've always pretended you don't care. But you've already tried to find him, and you failed, and I know very well why you failed even pursuing the answer through all the computers in the world."

Antony was trembling and totally lost. He knew that everything his cousin just said made sense, and something within him seemed to avoid the true answers.

Silence took over the conversation between the two cousins for a moment. Antony looked like he was trying to repress all the agony of the world inside his throat. He couldn't resist for long and asked Christine in a weak voice.

"What do you mean with all that?"

"Let me share something with you. Something that, in my opinion, should have been shown to you a long time ago."

Christine Sims shared with Antony a personal file from her computer, a file named *Hawaii holidays with my love.*

"Ok, Antony. I know you love to pry into my picture files, but you never checked this one before, probably because of the stupid name I gave it. But this name was just to evade you. I am sure you thought, 'hmmm… I don't wanna see pictures of my ugly cousin wearing a bikini.' Well, there aren't any holiday pictures here but answers, answers to your doubts."

A yellow file popped up on Antony's laptop screen as he trembled, sweating in silence.

"Enjoy your time checking the information inside it. Now I should go back to my work. Someone needs to work to keep the family business up. Goodbye."

The video call turned off, and Antony just saw the yellow file highlighted in the middle of the screen.

"She's lying. She's always lying," Antony whispered.

Suddenly, one of the cats meowed low, but Antony got startled as if he had just heard a lion roar.

"Ok, ok. Keep calm Antony. It's probably nothing."

Antony clicked the file and looked inside it, where there were several scanned documents from different hospitals and clinics. The first document made him even more intrigued because a different version of the Sims Corp symbol was stamped under it.

This symbol had never appeared this way since it had been patented by Allan Sims. Also, the documents made Antony very nervous, sweating more and more. He spent a long time just analyzing the first one and its symbol, experiencing an odd feeling that he had already seen it before. But then he stopped focusing on the symbol and started checking the other documents.

Antony's astonishment increased, especially when he realized that all the documents were medical reports. The first one showed the medical exams of a woman who had been through twenty-three abortions. Others were medical reports about uterine surgeries, artificial inseminations, drug detox, shock treatments, and different treatments involving a lot of antidepressants. Any human subjected to these kinds of treatments would probably be left with serious aftereffects or die. The most horrific part was that all these reports were under the name of the same patient, and they had been written more than twenty years ago.

"That's insane... and impossible too," Antony whispered.

Antony checked all the documents quickly, still thinking it was impossible for the same person to endure all these medical treatments.

Suddenly, the medical reports ended, and Antony found other kinds of scanned documents. These were contracts about the acquisition of expensive cars and huge properties around the world, all of them destined to the same woman who had been subjected to all those treatments in the previous documents.

"No, no, NOOOOO... That can't be it."

After all these contracts, Antony found a sequence of bizarre pictures he wished he had never seen. These pictures showed scenes of wild parties with young people completely drunk and high. Also, there were several pictures of orgies with people doing things Antony had only seen in porn. In every one of these pictures, a specific person appeared, always looking to be enjoying each moment — a teenage girl who was always wearing a silk mini skirt and tops stamped with different ancient heavy metal and punk rock bands. Her hair was long and black, sometimes with a blue or red lock. She looked crazy and seemed to be out of control — especially in the orgy ones. Her makeup was heavy, and on her arms, there were a lot of syringe marks. Antony recognized the face of this girl despite her younger appearance. The same girl whose name was in all the documents and medical reports.

The last thing in the file was a short black and white video recovered from a security camera inside an interrogatory room from a high angle. In this small room, there were just a simple chair and a white table where this girl from the pictures sat. She was sitting with her arms crossed and a long face the whole time. She had probably been arrested by the police from another crazy party.

Antony was trembling as he almost gave up on pressing play, but after taking a deep breath, he went on.

In the video, the girl called the security guard next to her and asked for a phone. The security guard handed her an old phone, which she grabbed angrily, like a caged animal, and then she made a call. Antony could hear only her voice, not the voice of whoever was on the other side of the call. Her voice sounded angry.

"Ok, ok. I'll sign anything you want. But I want the castle in Monaco, and your collection of Ferraris too."

"WHAT? The collection was a gift. Come on... give me at least five Ferraris and two Harleys. One day I'll learn how to ride this shit on the road."

"No, no... and the penthouse apartment in Brussels is mine. I hate this city, but it is a calm place to drink."

"Also, I want that penthouse apartment in Rio de Janeiro. I know you hate hot places like Brazil. And the business jet too, the big one with a pilot available on weekends."

"WHAT? Expensive? Do you think I'll sell my uterus cheap? You want an heir to your fucking corporation. If I'm the only one who can give it to you, I can make the price, ok?"

"Humm... Well, finally we have an agreement. So just send over the papers so I can sign them and tell someone to get me out of this shit hole. QUICK!"

The girl turned off the phone and dropped it on the table. Then she went back to crossing her arms, keeping her long face. She looked around the room, searching for something until she found the camera high up on the wall. She stared directly at the camera and gave it her middle finger. The video ended on this scene, a moment that would be etched in Antony's mind for a long time.

After checking all the papers, pictures, and the video, Antony was sure he would never be the same person again. He wouldn't be the good guy who was trying to save the world anymore because he recognized the girl he was seeing. He knew the name and the signature of the girl that was on the documents and the medical reports. It was Angela.

CHAPTER 52

The weather inside Cloud was controlled and stable all the time, every single day. Warm days and nights and some cold ones in the winter. People acted like nothing special was happening outside, definitely not the preparation for an upcoming war. Actually, nobody even imagined what was happening and how dangerous it would all become in a few hours.

In her office, Christine Sims was making the last modifications to the system of Cloud Base and locking all ways, or *gates,* as she preferred to call it, to enter Cloud's virtual world. Nobody would be able to get in or out while the war was going on outside — not even Antony. That was the only way Christine found to protect the business of her family and other future agreements. All these ideas worked for her and Yelena because they couldn't imagine what would be coming from the huge Black Hole some miles away from Cloud Base.

The moment to lock all *gates* was coming, but Christine left some of them open to allow Antony to enter and visit his mother one last time. Not because she thought Cloud could be destroyed, but because she knew Antony wouldn't see his mother again.

***

Angela was calm, just sweeping the floor of her home's living room. Everything in this part of the home — heavy furniture, sofas, carpets, little chandeliers — were in pastel shades, which varied from pink to white and green. There were pictures of the Sims Family in relaxed moments hung on every wall of the room, all together and smiling like a normal family. It was sunny outside, the artificial light created by Cloud's programmers came in through the windows and patterned curtains.

The moment Angela was sweeping close to the coffee table, she was startled by Antony's unexpected arrival in the middle of the room, holding in his hands several papers and pictures. He was trembling with fury, his face sweating, and his neck veins dilated to the point where they seemed like they would explode.

"Tony? What is going —"

"MONSTER! WHAT THE FUCK DID YOU DO WITH MY LIFE?"

"What… what do you mean? What is happening?" Angela was scared.

"What do I mean? I mean this."

Antony threw the papers violently at Angela's face. She lost her balance and almost fell back as all the papers and pictures spread over the living room.

When Angela saw all the documents and medical reports with her name on them and the pictures from when she was young, she got desperate. Tears rolled down her face as she sat down on the sofa, trying to understand what was happening in that moment, trying to understand how a forgotten past could come back like this — a past that she had strived every single day to forget. Each picture and paper that Angela saw on the sofa and the coffee table brought back terrible memories from a kind of person she didn't recognize anymore, but who she had been in the past, and there was no way to run away from this truth.

"How... how did you get all of these?"

"Does it matter? Does it matter how I got all that? This is the proof that you, your dad, and the rest of this damn family are rotten. THIS FAMILY DESERVES TO DIE. EVERYBODY."

"No, no, look, Tony I... was different but I changed, I changed when I saw you, my son."

"I'M NOT YOUR SON... look at these papers, I'm just an agreement. I was traded for luxurious cars and mansions. For what? To solve the problems of your father's damn big computer system. My life is a lie that everyone in the family always knew."

"YOU'RE MY SON... you were generated within me. I made a lot of mistakes, I know... but when I saw in my arms, I felt... I felt that everything could be different."

"No, no, no... when you saw me, you said: 'look at my new Ferrari' or 'look at my mansion in Monaco'. I saw the number of abortions in these papers! How many times did you try to kill me before I was born?"

"Sorry, I'm so sorry, but please, forgive me. Since you were born, I was never the same. Believe me."

She hugged herself and couldn't stop crying. *How to explain the inexplicable*, this was what was going through her mind as Antony kept walking nonstop around the living room, like a starved beast in a cage.

"Belief, trust, love... these words don't fit in with this family," Antony said loudly.

"That's why your grandfather wanted you… to give this family something good, someone who could give him hope."

"To me, he wanted someone who would be easy to manipulate, like you and Uncle Peter, days ago in this house, trying to convince me that creating this damn place was a good idea. I have never seen a video where grandfather didn't seem like a machine talking about business. I… I don't know why I keep calling him grandfather. He is fake. Everything he has created is fake, including this family."

"You didn't know him like I did; he was an amazing person, but he lost control of what he created. You are the proof that he still wanted to save this family. Oh… Lord, I'm so sorry."

"Lord, Jesus, none of these fairy tale characters are here to fix this situation. And how about this bullshit about your religion? Ah… Now, I understand. Nobody would suspect a loving catholic old woman. This is the best disguise to manipulate me."

"No, no, my love for you and for the Lord is real. These are my gifts, my great gifts that I received when you were born —"

"STOP, STOP saying that I was born. I wasn't born. I'm a damn laboratory experiment. I'm a… ah… a cluster of sperm inserted inside your rotten uterus that you negotiated with your damn father. That's all." Antony cried a lot, and tears covered his furious flushed face.

Angela finally got up slowly and used her last strength to take a deep breath. Despite this world being virtual, their consciousness could feel all the pain constantly. She stood firm, staring bravely at her son, as if those would be the last words she would ever say to him.

"Ok… so, what are you going to do? If you are sure you're just a damn experiment like you said. Shut me down then. If you don't believe in my love for you, the love that I have given you since you were a little baby in my arms, the love you think is fake... please, turn me off of this place and end the story of our family."

Antony stayed silent for a while, staring at his mother, his wide eyes in flames.

"Once... you told me you thought you were a messiah. I have never known a messiah acting this way with his mother."

By an animal instinct, he went up to Angela and caught her firmly by her collar with one hand as he lifted his other hand, ready to strike. He trembled, holding his mother's face in front of him. His other hand trembled as well, but the inevitable punch didn't happen. Suddenly Angela broke the silence in a calm but confident voice.

"Go on... my son, prove that you don't believe in my love. And prove you aren't a messiah once and for all."

Silence took over again.

The violent argument between mother and son was being witnessed by Christine, outside Cloud, from her comfortable office, through one of the screens on the top of her wide high-tech desk. She looked excited with a sarcastic smile on her face, watching Antony and Angela facing each other. Also, she was sure she had already got what she wanted.

"Well, I would love to watch the end of this *loving* family reunion, but I'm very busy today. Goodbye, my dear cousin and aunt."

Christine pressed a button on her desk softly, making a lot of little screens and dots of light appear at once. The top of the desk became a pool of lights and screens blinking like a psychedelic hallucination that she watched with impressed eyes. The desk lit her face up with different colors, creating a kind of horror scene where Christine felt for a moment, that she could control everything.

"The time is coming, and it could be the beginning of a new age," Christine whispered.

All *gates* between the outside and Cloud were locked immediately and only through Christine's office would it be possible to get in. This way, Cloud Base's general manager could protect and preserve the Sims Family empire during the upcoming war.

The consciousness of people who died from that moment on would be transferred to Cloud Base Two in Cheyenne. Christine had carefully closed all connections that people used to get in and visit Cloud, so just employees and powerful people close to the Sims Family had access to it in these last days. But from this moment, only Christine had access from her office, and only one person could break in, Antony. At any time, he could find a way to get in, turn Cloud off and destroy the business of the family. Despite Christine not believing that her cousin could have created Key, it was a possibility — and in the current situation, she couldn't risk it. Her plan was simple: destroy Antony's emotional stability and lock all gates while he was still inside Cloud.

In that moment, Antony was the only human from the outside connected into Cloud, and Christine knew very well that locking the gates with someone connected from the outside could cause dangerous mental health problems. She believed that if she cut off Antony's connection, he would be stuck inside Cloud through the chip installed in his brain, but she had forgotten a detail. A long time ago, Antony had disabled this chip, and that's what allowed him to escape from the last outrage of Uncle Peter in the mansion in California. Another possibility was that the brain wouldn't be able to stand such an abrupt change when the connection got cut off suddenly, and the user could be killed instantly, without a chance of being sent into Cloud. However, Christine wasn't worried about that.

In one second, Angela saw her son ready to punch her in the face, and in the next, he had simply vanished. She fell on the sofa and on her old pictures, crying intensely without knowing for sure whether she felt relieved or even more desperate.

Angela stayed there, unable to move, shedding bitter tears over the pictures which insisted on showing her disgusting past that had come back to destroy the rest of her life.

CHAPTER 53

Antony's eyes opened suddenly, and he felt like he had just escaped from drowning. His vision was blurry, and his head felt as if it was going to explode. He squirmed on the floor of the jet, where he had been the whole time since that terrible meeting with his mother.

The jet continued flying toward Cloud Base, approaching the desert and piloted by the AI.

Antony finally regained some strength a little, but just enough to remove the Headgate from his forehead. The three moggie cats looked scared, watching him sit down on the floor in the corner of the passenger cabin — trembling and sweating, his eyes tearing up. He looked like a lost child abandoned by his parents in an orphanage, but with an expression akin to that of a drug addict without his drugs.

One of the cats approached him and lay down on his lap, which was crossed. Antony caressed this cat softly as the other two approached as well and stayed with him.

"Hey, my friends. You are my friends, aren't you? You'd never abandon me. You'd never betray me… I know that."

Antony was completely out of his mind when suddenly the Jet's AI rang out with information about the flight.

"Young Mr. Sims, in twenty minutes, we'll arrive at Cloud Base."

Antony stayed silent for a moment, just caressing the cats. The flight was calm, and outside, the afternoon was ending when he gave the AI a horrible order in a weak and trembling voice.

"AI?"

"At your service, Young Mr. Sims."

"I want... I want you to turn off the jet's engines and let it fall on the desert... and explode."

"Can you repeat, Young Mr. Sims?"

"Turn off this damn jet and let it explode on the desert down there."

"Young Mr. Sims, I can't accept an order that will endanger your life."

"Just do it quickly, AI," Antony looked as if he were going to explode at any time.

"Young Mr. Sims, according to the security protocols I must disobey —"

"FUCK IT, ARE MY ORDERS WORTH NOTHING HERE? I'M ANTONY SIMS, THE HEIR OF SIMS CORPORATION — THE FUCKING RICHEST CORP IN THE WORLD — AND I CAN'T COMMAND MY OWN JET. I WANT YOU TO TURN OFF THIS DAMN JET RIGHT NOW AND END MY LIFE."

Antony's shouts frightened the cats that ran away from him and started to meow desperately. Antony changed his expression immediately when he realized he was making the cats afraid.

He felt an intense agony as he crawled onto the floor with his hands outstretched, trying to call the cats close to him again.

"No, no, no… sorry. I'm so sorry… you're my best friends. Please, please come back. Please!"

The cats slowly came back close to him, and Antony caressed them carefully again, his face filled with sadness the whole time.

"Yes… come on. You're my friends. My best friends. You can't leave me alone. You can't… I… I won't let anything bad happen to you. I promise… I promise," he hugged one of the cats, unable to repress his desperation.

"Young Mr. Sims, according to the security protocols, I must disobey any orders that endanger your life. Please, review your last order. If you don't review it, I'll have to follow the previous order, which was to keep the trajectory to Cloud Base."

"…"

"Young Mr. Sims?"

"So… AI… get away from Cloud Base —"

"Young Mr. Sims?"

"Get away from Cloud Base and locate the next abandoned airstrip or hangar. Then just land there. I don't care where it is, as long as there is nobody there… I don't care about the world… not anymore."

"Ok, Young Mr. Sims."

Antony stayed sitting down in the corner of the passenger cabin with the moggie cats, like an abandoned child expecting nothing.

CHAPTER 54

Life continued happily and without worries inside Cloud. The people inside enjoyed the work they loved, indulged in their favorite hobbies, and spent time with their loved ones in a world without violence, natural disasters, or diseases.

A true heaven, created by Allan Sims decades before, had changed the concepts of religion, *life after death*, superior entities and fear. The fear of death disappeared, and faith in religions and eternal salvation went with it. Allan Sims showed humanity that heaven was real, palpable, and no one had to follow any religious leaders and philosophies that only showed people an abstract idea of what we might find after death. Not anymore. Heaven was real.

Most people saw Cloud as a place of comfortable and well-deserved retirement. Companies used it to make money, selling the best places in there, and bad politicians always found a way to send people they didn't like to Cloud and keep them there forever — some of these were bad people, but most of them were innocents. Cloud had changed what had been the most important and vulnerable part of humanity, our imagination. For centuries, politicians had used the education system and media to manipulate it according to their interests, but Allan Sims' inventions had broken these strategies as the new concepts of religion and *life after death* overcame the manipulation through education. However, not even Allan Sims could have predicted that creatures from another world would come over and force the powerful people of Earth to stop trying to control the lives of others.

While billions of people's minds lived happily inside Cloud, around Cloud Base, a huge military organization was ready to counter the attack of anything that could come out of the great Black Hole, only a few miles away up in the sky.

It was a Sunday morning, seven days after the last battle, and, according to Muriel, the Angels' attacks would happen every seven days, each more violent than the last.

Despite the weather in the middle of the Nebraska desert, the sky was totally dark, like a starless midnight, with only a Black Hole that seemed more alive in the last few hours. It was spinning faster as lightning flashed intensely out of it, while thousands of soldiers, tanks, cannons, and giant robots stood in front of Cloud Base. They were positioned in line among big piles of wreckage, the remains of roads and train tracks, and long metal beams — most of them taller than the giant robots. These big piles were a part of their battle strategy; they would provide protection to soldiers and robots while they were in combat.

Everyone seemed to be apprehensive, looking up and expecting the worst possible outcome. Hundreds of combat drones, controlled by operators from remote terminals far away, were flying around the area like an immense swarm of iron wasps from the soldiers' point of view on the ground.

In front of the army, a giant robot armed with missiles on its shoulders and a long sword at its waist walked firmly and slowly. Its pilot was checking if the soldiers' and robots' positions were right for the start of the upcoming battle. This pilot was Glenda, and she was sure this checking was useless, but she didn't have any other option except trying to protect as many soldiers as she could and wishing with all her strength that everything might end well, allowing her to see her son safe again. However, she knew that it was almost impossible.

The robots, bullets, missiles, and the soldiers' vests were made out of the metal Antony had created based on the piece of Muriel's armor some days before in Glenda's Dockbot. Christine realized in the last battle that Antony's giant robots and their weapons were more resistant against the Angels' strikes. Therefore, she was clever enough to steal the sequence of metals that her cousin had used to build the giant robots and replicate this new kind of metal in her weapons, which she named Brave Metal.

Glenda stopped for a moment and noticed strange storm sounds coming from the Black Hole, which had started to move faster. The time was coming, and following an inexplicable instinct, she looked up toward Christine's office and saw something that made her very angry.

"Damn cow. You're likely thousands of miles away from here along with that damned Prime Minister."

Ten cameras fixed on tripods were set at the wide window of Christine's office. The general manager of Cloud Base would watch the war from somewhere safe, probably a military base nearby, along with Yelena Orlyk. Only a few programmers and many well-armed soldiers stayed inside Cloud Base, just in case some Angel got into the building.

On the other side of the Black Hole, Cassiel stood on the peak of a mountain watching the Black Hole spin some yards away from him. Suddenly, he closed his eyes like a priest ready to pray and whispered:

"The time is now. I feel my God's orders touching my ears."

Cassiel took his long copper hunting horn to his mouth and blew it, making a terribly loud bass sound — a mix of a lion's roar and thunder, so loud that it terrified everyone outside the Black Hole. *This is the sound of the end of the world*; that's what all soldiers and pilots thought at the same time. This sound echoed for a few seconds in the middle of the Nebraska desert, serving as an announcement for a brutal battle on Earth.

Every soldier and pilot focused their eyes on the Black Hole that looked like an immensely bright light, a purple hurricane that seemed to have a will of its own, swirling around in the sky.

Flashes, thunder, and fear. Suddenly, thousands of what looked like small dots came out of it. Some soldiers were trembling; others wished they hadn't agreed to stay there. *The end is near*, that was the quote that finally made sense in that moment to every man and woman on the ground.

An immense crowd of Angels flew over Cassiel, who just watched through his serene eyes all kinds of winged warriors on a gray sky crossing the Black Hole.

On the other side, the soldiers and pilots couldn't describe what they were seeing — a flock of eagles, a sandstorm, or a swarm of countless winged creatures. Fear and tension made the reasoning of the soldiers useless; all that remained was the survival instinct and the will to fight.

"Oh, Lord. It's a good opportunity for you to appear," Glenda whispered.

All kinds of giant Angels landed on the ground along with others that looked like humans. Every soldier and pilot were mesmerized when they saw the amount and variety of creatures — giants wearing armors of different colors: gold, copper, iron, ruby, and jade. Masks of different formats that resembled the faces of dragons, jackals, tigers, skulls, kabuki, goats, simply deformed helmets, and other bizarre ones. They wielded large medieval weapons like swords, morning stars, spears, shields, daggers, axes, and hammers. Every one of them looked demonic and ready to destroy anything they saw in front of them. Soldiers and pilots stayed frozen, just watching the horizon being filled with bizarre giant Angels that made it look like a costume party to celebrate the end of the world. Other Angels the same size as humans were positioned among the giant ones, and they were just as diverse in their armors and weapons. Everyone had their wings open, wielding their medieval weapons and with bloodshot eyes.

Other Angels stayed up in the sky, flying in circles, just watching the battlefield and the armed drones that flew over Cloud Base. Muriel and Abaddon flew alongside these Angels, just waiting for the next orders from Cassiel on the other side of the Black Hole.

Muriel was impressed with the humans' capacity to organize so many warriors and robots in such a short time. It made her think that they could beat the first group of Angels, but not the second.

Glenda took a deep breath and arranged her headset. She seemed to swallow all kinds of fears that existed in the universe with each giant Angel that landed in front of the troops. As they landed, they drew their weapons and readied themselves for attack. Swords, spears, shields, morning stars, clubs, hammers, axes — all kinds of medieval weapons were in the hands of the armored Angels, while the humans had firearms, missiles, and cannons. Glenda insisted that Christine authorized the production of swords and daggers as well. The manager of Cloud base was confident that bullets and missiles could guarantee their victory, but if all this artillery ended, only the brave who kept fighting with swords and daggers would remain. With this in mind, Christine named this kind of metal Brave Metal — a tribute to those who would accept to die for a poor home in Cloud, according to Glenda.

Thousands of Angels on the ground positioned themselves to attack, covering an area bigger than the Black Hole, while in the sky, several Angels were flying in circles.

Only a few yards separated the Angel troops from the human troops, which seemed to be in higher numbers, calming Glenda down a little, but not enough. She turned on her headset and tried to hide her nervous voice, but it didn't matter to the soldiers who were there.

"So, bros... I mean soldiers. Men and women ready to save this country and the world too. Today can be a good day to live... or a good day to die... Just fight for the first option."

Thousands of clicks from the guns, missile launches, tanks, and cannons being unlocked echoed through the lines of the human troops, making their hearts beat faster and their instincts sharpen. Unlike the

Angel troops, that looked like an army made of frozen warriors, just wielding their weapons.

"Come on, crazy Angels. Just let the fight begin," Glenda said out loud.

The Angels stopped coming, and from the other side of the Black Hole, Cassiel blew his copper hunting horn again, and another bass sound echoed through the skies of the two worlds. This was the signal for the Angels to attack.

On land, the winged creatures marched forward ferociously, screaming, a sharp and bizarre sound like that of angry eagles echoed through the desert.

In the sky, the attack was the same. The Angels, led by Muriel and Abaddon, wielding their swords, bravely attacked the combat drones that shot their machine guns and launched missiles at them with no mercy, making the dark sky shine with countless explosions — like fireworks on New Year's Eve, but celebrating a terrible war.

Some Angels quickly dodged the shots and missiles; others were hit hard and fell, injured by the power of the ammunition made out of Brave Metal. Some of these Angels fell, while others still had strength to go back to battle.

Muriel was the fastest and most agile warrior in the sky, keeping focus the whole time and striking each combat drone precisely on their wings. Her wings moved wildly, but her breath didn't change, just like her cold eyes. Each stroke left behind a damaged drone or pieces of them on fire, falling, spiraling down to the ground. However, she was impressed by how resistant these machines were, built out of the same metal as her armor. She realized, in practice, that she had to use her strength beyond normal to shoot them down.

"The humans evolved since the last battle. They learned fast," Muriel whispered.

On the ground, Glenda drew her sword and ran toward the Angels, with a furious shout that pushed all pilots to scream as well, creating a long wave of giant robots going fast — like the barbarian troops of the medieval age.

Hundreds of soldiers went on bravely too, running among the robots, shooting their machine guns while others stayed on top of piles, shooting their bazookas and rifles. Near the building, tanks shot nonstop toward the Angel troops, hitting them hard, until pieces of their armors and wings were launched through the air. Also, the giant robots shooting missiles from the launchers on their shoulders hit the crowd of Angels that was coming towards them hard. These Angels fell before the power of human artillery, an artillery made of Brave Metal, and they were stunned. It allowed the giant robots to slash their opponents more easily, but despite the humans' good strategy, the Angels were hard to kill and fought bravely.

The sounds of explosions and metal clanging echoed through the dark desert like a devilish symphony. In that moment, there was no more reason, only survival instincts and a strong will to kill.

Glenda was the most experienced and fastest pilot on the battlefield, thanks to her new robot, more resistant, better armed, and easy to control. The right technology of Sims Corp built the best machines to fight against the Angels, an evolution from what Antony Sims had created by himself in the simple Dockbot in Glenda's community.

The main general of the human troops struck hard and fast all Angels who showed up in front of her. A golden one was wearing a mask that looked like a goat, another wore a red armor with a dragon mask, and the following, a green one, was wearing a helmet that looked like a hokey mask. Each more bizarre than the last, but Glenda didn't have time to reason, only to slash and launch missiles.

The battle went on furiously around Cloud Base. Giant robots, Angels, soldiers, drones — all warriors were using their weapons bravely, showing no mercy, on the land and up in the sky. For a while, the building was intact like an immense white heart in the middle of a dark desert, and the population inside Cloud didn't even imagine what was happening outside. However, it wouldn't stay that way for long.

Glenda struck down more and more opponents without even blinking, no time to take a breath. Suddenly, for a short moment, many Angels surrounded her and positioned themselves, just waiting for the best moment to attack. In this brief moment, Glenda realized the terrible situation she was in. Drones in flames, falling from the sky, giant robots in flames being destroyed by Angel strokes, soldier, bloodied, soldiers crawling on the ground, some screaming while others no longer screamed. Shots, missiles, bombs, explosions, cries, courage, and fear. In that short moment, Glenda got to breathe and say:

"Where are the people responsible for this fucking situation?"

CHAPTER 55

It was a normal sunny Sunday morning in a vast forest located between Wisconsin and Iowa. This forest's appearance seemed calm, quiet, as always. The noisiest thing there was a fresh breeze that moved some dry twigs. But suddenly, the sky became dark, not because of some storm or eclipse, but because an immense fleet of fighter aircraft, strategic bombers, and combat drones were crossing the sky toward the Nebraska desert. This fleet was so massive that it was able to stop sunlight from reaching wide areas below, turning day into night. This was the American Air Force.

These machines had just left the Spartan Military Base in Wisconsin, and they were heading to what seemed like the last battle in human history.

Some pilots had already received the news that the battle around Cloud Base had just started, and that they had to be fast enough to arrive and stop a bigger tragedy. However, nobody on Earth knew that the Angels had a second group ready to attack, bigger than this fleet — much bigger.

One of the pilots, codenamed Eagle 8, piloting an F-35 Lighting V, was checking the communication system that connected all airplanes until he got in contact with the pilot codenamed Eagle 23, who was piloting an F-15 Strike Eagle III.

"Eagle 23, this is Eagle 8. Just checking the communication system and trajectory coordinates. Do you copy?"

"…"

"Eagle 23, do you copy?"

"This is Eagle 23. Communication system and trajectory coordinates just checked. All ok. Do you copy Eagle 8?"

"Eagle 23?"

"Eagle 8, listening."

The pilot in the F-35 Lighting V took a little while checking something on his board computer before he reestablished contact.

"Eagle 23, who are you?"

"Eagle 8, what do you mean?" The pilot's voice was calm and confident.

"Eagle 23, I'm checking your database, and I've just noticed that we've already been on several missions and trainings together."

"Eagle 8, what's the matter?"

"Eagle 23, the matter is your voice. I know the pilot that I'm checking on the database, and I'm sure that I'm not talking to the same person."

"Eagle 8, I know that. But I also know that my voice sounds familiar to you. Do you copy?"

"Eagle 23... yes, your voice sounds familiar," the pilot sounded confused, "but I can't remember where I've heard it before, yet."

"Maybe in Spartan Military Base. Do you copy?"

"Eagle 23, I was working with many soldiers and... no... it's impossible."

The pilot was mesmerized. He'd just realized something unthinkable about the voice he was hearing, and then, after taking a deep breath, he got the courage to ask.

"President Sowell?"

"Nice to meet you, Eagle 8."

"You shouldn't be in this mission, Mr. President. I must report it to my supreme commander —"

"Eagle 8, I guarantee that all the commanders you report to are here flying with us or in combat in Cloud Base. And if you report my presence here, the most they can do is report it to the president of the United States. And that's me."

"Sorry, Mr. President. It's an unusual situation for me."

"It's an unusual situation for everyone here, son. No worries. I hope you understand me, Eagle 8, but I have to stay where my people need me and where my soldiers need me."

"Yes. I understand. It's an honor flying with you, Mr. President... in the name of our country and our world."

"Thank you so much, Eagle 8. So, can you tell me how long it will be until we get to our destination?"

"About forty minutes, Mr. President."

"Ok, we still have time. Eagle 8, do you copy?"

"Yes, Mr. President."

"Can you keep my presence in this mission a secret for a while?"

"Of course, Mr.... I mean, Eagle 23."

"Excellent. Keep up the good work, Eagle 8. Over and out."

The immense fleet kept its imposing flight projecting a huge shadow on the land below, moving forward, full of uncertainties and fears.

CHAPTER 56

With a high kick, Glenda hit a giant Angel wearing silver armor with a helmet that looked like a buffalo. Then, she struck with her sword another one with a metal armor and a helmet in the shape of a lizard's head, right on his chest. This Angel tried to get away from Glenda and take the sword out of his chest, but his opponent was fast, and she launched a missile directly at his head, throwing his helmet far away as his body fell to the ground lifeless. Then she took her sword out from the Angel's chest, and with a fast movement, she hit three other Angels who were flying around her — launching arrows with their bows. Pieces of Angels' bodies fell down, and Glenda's sword was soaked in blood and feathers.

Glenda noticed three more giant Angels coming toward her, and before she lifted her sword again, several missiles and rockets were launched from behind her. She turned back for a moment and noticed many soldiers armed with bazookas and tanks covering her. But it didn't relieve the stress that had completely taken over her mind and body.

When every human in the battle thought that the situation couldn't get worse, the Black Hole roared one more time as if announcing one more terrible surprise.

Two giant armored Angels came out of the Black Hole, and all the soldiers stopped for a moment to watch them. They were like the others giant Angels, but their size was impressive. They were about four hundred feet tall, four times taller than the other giants. One of them wore silver and red armor and a medieval helmet that looked like an eagle's head, while the other wore a golden and metal armor that resembled a sphinx. Both were armed with spears shaped like a long screw.

They flew in circles, seeming to take over the whole sky with their long white wings until they landed heavily, making the ground shake for a moment — like the beginning of an earthquake. The battle continued, unstoppable and violent as these two new giant Angels just stood, watching the carnage around them and calculating their next moves — like two enormous ice statues.

Glenda had started this battle with some hope, but after seeing these new opponents, it had vanished completely. In her tired voice, she said to herself:

"Where are you, Antony? Now's a good time for you to bring us a miracle."

CHAPTER 57

A fresh breeze blew among the pines in a calm forest where just the birds' singing broke the silence and an almost inaudible faraway bass sound too.

In a large dirt path that crossed this forest, an old redneck walked slowly, as if one day lasted forty-eight hours. This solitary man looked kind and always had a likable smile on his face. He was in his eighties and was wearing worn-out denim overalls and a plaid shirt. His heavy body didn't seem to feel its age as he easily carried a heavy old fishing tackle box in one hand and some fishing rods and an iron bucket in the other. His round face exhibited a short white beard, a childish smile, and eyes which seemed in awe of everything around him, like butterflies and birds. A worn-out straw hat protected his head from the first morning light.

This man had kept the same routine for decades every single weekend, but he admired every detail around his path as if he were walking through it for the first time. The bushes wet with dew, the squirrels climbing up pine trees, the combination of the scents of flowers and mud that was spread through the air. Every step for him was a new and honored chance to contemplate nature. But suddenly he noticed something very unusual.

He walked next to an old, abandoned wood hangar as he usually did. A hangar that had been almost completely hidden by bushes, pines, and dry leaves for a long time. A place well known to this old man despite something big right in front of it. A dark blue jet with its door wide open.

The old man stopped for a moment, watched mesmerized by the very modern machine in front of him, and said to himself:

"Humm, that's a beautiful machine!"

Then he kept on his customary walk through the path.

After a few minutes walking, the old man noticed another surprise at the pier on the river where he had fished every weekend. A thin young man was sitting down on the edge of the pier, just looking at the pines and blue hills on the horizon.

The old man took a minute just to watch the crystal-clear water of the huge lake and the unexpected visitor who looked very sad. Then he confidently walked toward the young man until he was at his side and started a conversation.

"Hey, boy, you forgot your fishing rod, I guess," this man always kept a friendly smile.

Antony was a little startled by the unexpectable presence, but not enough to get up and change his sad expression.

"Oh, no... sorry, I didn't come here to fish. I... I'm just where I have to be, I guess."

"Hehehe, wise words, boy. Well, you seem to want to be alone. I can look for another place to go fishing."

"No, please," Antony's voice sounded weak, "I'm the stranger here. You seem to have been fishing here for a long time."

"Hmmm... yes, I guess about forty years. I fish here every weekend."

"Forty years? Wow... I'm very sorry for disturbing your routine."

"No worries, boy. It's good to see someone different here."

"You probably see many people come to this beautiful place."

"Hmmm... in forty years... I think you're the first."

"Wow."

For a moment, Antony wanted to feel pity for this man, but the pain inside him was so deep, and the smile of the old man was so frank that it was difficult to choose his following words.

"Ok, boy, if you don't mind, I'm gonna take my seat here. Well, a nice day for fishing, huh."

The old man took a seat on edge of the pier by Antony's side and got his stuff ready while he talked friendly.

Antony kept sitting down, just looking at the man who was happy beyond normal. However, anyone would seem happier than Antony.

"Hey, you came on the blue machine back there? It's a beautiful airplane. Are you an *airplane pilot*?"

Antony thought for a while before explaining about his jet, trying to find the easiest way to talk to this simple man.

"Oh, it's not an airplane. It's a jet and... anyway. Yes, it's my airplane. There's a computer inside it... This computer pilots it, and it brought me here."

"It's yours? You must be very rich to have your own airplane. If I had one of those, I wouldn't look so sad... hehehe... sorry, I'm just joking. If I catch a big fish, we can cook it and eat together. What do you think?"

Antony looked doubtful about this man's kindness, especially after all his family and friends had turned against him. Another thought that arose suddenly in his mind was that he was the most known person in the world, because his face had been in all TV shows and internet news websites every single day in the last few years — and this old man seemed to have never seen him before.

"Sorry... I'm not a big fan of seafood and —"

"Oh, I almost forgot to introduce myself. That's rather impolite."

The old man reached out his hand for a handshake.

"My name is George Owens Davis, but you can call me George. And you, boy?"

"I'm Antony... Antony Sims."

"Tony Sims? Sounds like the name of a superhero hehehe... Nice to meet you, Antony."

"Oh... nice to meet you, George. But I'm not a hero or anyone special."

They shook hands, George excited and Antony completely downcast. In that moment, Antony remembered the conversations he had with Nivedita and how he could be someone special, like a new messiah or someone who could fix the world's problems.

"I'm not special either, and to be honest, I don't know what it means hehehe... Anyway, feel free to take a fishing rod, boy. I have more here. It's amazing fishing while talking to someone. I can give you some tips to catch the big ones."

"Well... thanks, but I prefer to enjoy this landscape. It looks like a good place to hear my own thoughts. It is like a... a —"

"Like the desert where the Lord was tempted by the devil, but more beautiful hehehe... I have felt this silence too for the last decades, and I love it. I don't know why I remembered this now. Maybe because I haven't talked to anyone in a long time."

"Yes, when the Messiah was in the desert until the devil appeared and offered Him all the kingdoms in the world, then He refused it. Nivedi... I mean... a close friend told me this story once," Antony looked interested in the conversation but still sad.

"Hmmm, you look like a wise boy. You're right. Yeh, I know you don't have all the kingdoms in the world hehehe… but I noticed you're rich and nobody can tempt you, I guess."

"Oh, I've never thought about it. How could someone tempt me? And tempt me with what? A rich boy with a… an airplane, as you said," Antony looked doubtful.

"Hmmm, if you aren't happy, there is something more than money or airplanes that you don't have yet. Anyway. Are you sure that you don't wanna fish?"

"Oh, maybe. I can try."

"Please, take a fishing rod. The bait is already on the hook."

Antony took the fishing rod offered by George and threw the fish hook into the water.

They looked like grandfather and grandson fishing together on a happy and calm Sunday morning, although, to Antony, there wasn't any more happiness or family members. For a moment, watching the reflection of his feet on the lake's water, he remembered Nivedita's teachings again and thought over the words George had just said. *The devil offered all kingdoms in the world to the Messiah, and I was born already owning a huge part of the world. Nobody offered me power and kingdoms. I was already born with all that, and here I am, an empty person unable to help the world that he was given on a plate.* That's what flowed through his mind.

Antony knew very well that he had been born fated to be someone great enough to influence and maybe control the course of the world, but he hadn't wanted it, he wanted to help the world instead. At this moment, the world didn't seem to want his help — and his family had confirmed it.

Suddenly, a noise coming from the water interrupted Antony's thoughts, and George's friendly voice got his attention.

"Hey, look at that! It's a black crappie. They love this kind of clear water hehehe... I hope to get some bigger than this one," George had just caught his first fish of the day.

"Oh... nice!"

"Hey boy, when you were coming over in your airplane, did you see what is happening over there?"

"Where?"

George put the fish in the iron bucket on his side, then he pointed to the left side of the horizon.

"There, in the dark part of the horizon. It appeared about two weeks ago. Now you're here. Maybe you flew over there, I guess. I had never seen it before."

Antony looked at the horizon between the last green hills covered in pine trees, huge fields, and the blue skies above. There was a small part of the horizon that was totally dark with several red dots blinking nonstop. A detail that almost went unnoticed on the landscape in front of him, and that he wouldn't have seen if George hadn't shown it to him. Also, Antony noticed a bass sound coming from this part of the horizon that sounded like big explosions in the distance.

Only then had Antony realized that his jet hadn't landed so far from the desert where Cloud Base was located, and he would have a panoramic view of the beginning of the end of the world.

"Do you know what could be happening there, boy?"

"Well," Antony took a deep breath, "It's a battle between Angels and humans. Actually, it's the beginning of the Apocalypse. Shortly, this battle will turn into a grand war, and it will destroy all life on Earth. Long story short... we are on borrowed time on this planet."

Antony looked at George, who seemed scared, which made him think that he had been too rough with his words. But after a few seconds, George laughed a lot, almost letting himself fall from the pier.

"HAHAHA... You're a funny boy... very funny. You almost got me HAHAHA... Hey, look at that, it's a channel catfish. It's small, but nice," George caught another fish.

"Are you sure you don't know me?" Antony looked suspicious.

"No... Why?

"Nothing. It's ok."

The simplicity of the old man impressed Antony, who turned his eyes away from the dark horizon. Probably because he had never met such a simple person like this old man, despite the people he had met in Glenda's community. But this man was different. He looked incapable of being a bad person.

"Sorry, boy, but you look like you should be somewhere else, I guess."

"No, I shouldn't... I am where I have to be. That's it."

"Wise words. Well, so let's go fishing before the world ends hehehe... sorry."

"No worries and... yes, let's fish."

Antony kept his focus on the fish hook floating in the water... water that showed his distorted reflection.

CHAPTER 58

Explosions, shots, desperate screams in a dark landscape made out of the wreckage of bridges and roads. The explosions' fire spread hot winds that little by little diminished the soldiers' will to stay alive. Many of them just decided to let some winged creature kill them, and they were taken directly to a comfortable place in Cloud Base Two. In the beginning of the battle, it was easy to believe and keep their courage, but in practice and against these terrible creatures, their courage was easily depleted.

Screams of pain, screams pushing to attack, screams asking for help. All these screams mixed with the sounds of the metal from the Angels' armors and giant robots clashing.

Shots, missiles, bombs, the clanging of swords, and the buzz of arrows.

The ground was already covered in corpses, both of Angels and soldiers, the wreckage of giant robots, forgotten guns, and medieval weapons. Also, blood traces looked like veins on the injured land.

This was the scenery that Muriel saw from up in the dark sky. Her face was not as calm as she had shown it to be in the last challenges against the army, in Paris, and against Abaddon inside Cloud Base. She felt that time was passing, and her whole mission with Antony since they were kids seemed to be wasted.

Muriel decided to leave her position in the battle in the sky and went toward the Black Hole. She flew graciously with vivid movements of her silver wings and crossed the huge gate that seemed to pulse up in the sky. On the other side, she flew over through another gray sky, below her there were mountain peaks that arose from an immense ocean of fog. A landscape she had flown over along with Abaddon a few days ago.

On one of these peaks, Muriel caught sight of Cassiel holding his copper hunting horn. Despite being focused on the horizon, he felt her approaching.

Muriel looked very worried, breathing heavily, when she landed next to him.

"Cassiel... my master, please... we need more time —"

"Time? For what?" Cassiel looked very serious.

"To stop the apocalypse. These two worlds can be saved. Please... I still believe —"

"It does not matter what you believe," Cassiel said in a severe tone, "you do not need to believe anything. You did your mission and now you must wait and see the consequences of that. That is all. What will happen depends only on our God."

Tears started rolling down Muriel's face, but she stayed serious. She was trembling, looking at her master, not sure what she could say.

"And... if I failed?"

"Failure is a creation of stupid humans and their egocentric expectations. They think they can control everything and when they cannot, they call it a failure. Failure is an irrational and useless attempt to predict or control the nature of the universe. Don't imitate the humans! You are better than them. You did what you thought had to be done. The consequences of what you did, only our God can decide. Doubting your actions is doubting our God... and a universe that cannot be controlled."

"But... my Messiah disappeared, and time is ending. I cannot teach him anything else or look for him."

Cassiel approached her and slowly touched her face. His hand was white and wrinkled, but his gesture was careful. He wiped Muriel's tears with the tips of his fingers. From this moment on, he talked in a calmer tone.

"Time? How long were you in your mission?"

"All my life."

"If you gave your life to this mission, a life will be necessary to complete this mission. Our God knows how and where everything must happen. In this life or in the next."

"Yes... our God knows."

"I never told you that but... you remind me of Him."

"Remind you of whom? I do not understand, master."

"Our God... you are like Him... Our God wants the humans to learn how to become better beings by themselves. That is the matter... If only one human, at least one human, destroys The Cemetery of Souls, that means the humans have finally learned it, and this one can lead humanity to a path of goodness. Thus, our warriors will be obligated to stop the war —"

"But if our warriors destroy Cloud, then they will be obligated to kill the rest of humanity, because that is the proof that humans cannot learn by themselves... I know... but to me, that sounds unfair," Muriel sounded hopeless.

"I know you studied the other messiahs who lived on Earth... and you know that humanity only changed centuries after these messiahs left the Earth."

"To me, it still sounds unfair —"

"But it is the way our God wants humanity to carry on... Now you must go back to battle. Our people need you."

"Yes, my Master. I am going."

Muriel left Cassiel and flew toward the Black Hole. She felt as if the conversation had been useless, and her mission had reached a point of becoming a waste of her life. However, at the same time, she believed in her God and in the words of her wise master.

Cassiel stayed in the same place, just waiting for the next command from his God, watching a horizon completely full of Angels' silhouettes — so full that the horizon seemed to have been taken over by a swarm of dark insects, so full that Cassiel's eyes couldn't see anything else anywhere else in the horizons, except these thousands of Angels — Angels ready to attack Earth at his command.

CHAPTER 59

"Yeh, one more catfish. This is the big one hehehe."

"Oh... you're very lucky. I have got nothing so far, and you got more than ten fish."

Antony was still sad as George laughed with several fish inside the iron bucket at his side.

"Maybe my bait isn't as good as yours."

"It's the same bait, boy. The true bait is in your will. Or... I'd better say... in your mind. Yes, in your mind."

"In my mind? What do you mean?" Antony looked confused.

"I mean... you gave me a lot of ideas. Maybe because for a long time I have had no one to talk to, I guess. Anyway, I mean the bait is just part of what attracts the fish. We have to have a will to fish, a true will to get what we wish. Our wish is already in our mind. It depends only on us to get it done. If you don't have a true will to do what you want, you can have the best bait in the world and not get any fish. Also, the fish know that."

"You say that as if the fish want to be caught by us. Like they are suicidal."

"Hehehe... this is the circle of life. I mean... there are the hunters and the prey, the weak and the strong. Everything in the world is different, but it all works in amazing harmony. We can say the fish know that their mission is to swim through the water, be hunted, and become the meal for other animals, including us, of course. Everything works well in nature because everyone knows perfectly what their role in the world is. Even if their role is to die."

"Wow, but that's confusing to me. Being born just to die, and knowing that."

"Hey, look at the landscape in front of you, look at the birds in the sky and the flowers growing in the fields. They don't labor more than necessary, more than they can. They are what they have to be and do what has to be done. Unfortunately, humans showed up and insisted on taking control of all of this. Poor people," George looked a little sad watching the horizon. "Look at these fish. They know that their fate is to be hunted and feed other animals... for them to survive, then other animals will do the same. Because this is their role. Look at this place, my whole life I have seen this landscape, and I can't understand why people would change any of it. Death... death is sad and painful, but it's a part of us. If we couldn't die, we wouldn't be humans, but something else. I don't know if we would be something good or bad. But certainly, we couldn't be called humans."

"Wow... but I'm thinking about this now. Animals, trees, all kinds of living beings in nature, like these fish, know well what their roles are, but they don't have the capacity to create something more like us, humans."

"Humm... can you go on?"

"I mean... if we were born with the capacity to transform nature, that's probably our role. We were born to modify the world, to control nature. If we hadn't done it, well, we'd still be living in caves and running away from wild animals, I guess."

"Wise words, boy. Also, I think there is one more role humans have besides controlling the world: controlling ourselves. Better yet... knowing ourselves. Fish have the capacity to swim away and die alone, vanishing into the bottom of the lake. But they prefer to be hunted and feed others. So does humanity have the capacity to control the world, to destroy it and vanish from their own history, but they prefer to live here, have a family, and take care of each other... for a time, I guess. My doubts are: how much can we control the world? And when can we just live in peace with our world? Just in peace... because for me, living in peace is our true role."

"Yes, but humans seem, you know, to be afraid of being in peace and always prefer to create their own chaos... no, sorry... it's just a crazy idea that came to my mind now."

"No, you're right, boy," George looked sad for the first time. "Humanity has always tried to control what they fear... we're the only animals that want to control our own fears. But we still haven't realized that what we truly fear is to just live by ourselves... and die by ourselves. Anyway... I think you're tired of listening to the laments of an old man. Suddenly I just felt like blurting it all out... sorry, boy."

George seemed to just then wake up from a trance as Antony got emotional. He tried to change the subject so he wouldn't see this old man sad again.

"No, no worries. You're an amazing person. I love this kind of conversation. Wow... and... I noticed, when I landed my jet... I mean airplane, some big houses destroyed by time, like the hangar back there. This was once a very nice place, I guess."

"Nice place? Hehehe... this place was amazing, full of fun and happy people," George recovered his cheerful attitude.

"Wow... sorry, I'm the stranger here, and you asked nothing about my life. That is very respectful. But I would like to know more about my... my *host*. Your family or if you have kids. If you want to talk about it, of course."

"Hehehe... Like you just said, 'I'm just where I have to be.' I think the correct would be 'We're just where we have to be, doesn't matter if our actions were good or bad. These actions always lead us to do the right things in the right place'... Our past doesn't really matter. We're just here to contemplate this amazing landscape. But don't worry about asking anything. Well, like I was saying, this place was amazing in the past. Shops, bars, parks, parties, baseball games — a lot of people used to come here to have fun. I had a beautiful wife, Mary, and eighteen kids —"

"Eighteen kids? Wow, you're a baby making-machine, if you know what I mean."

"No, I don't get it."

"Oh... Ok. Go on."

"Of course, all these kids weren't mine. Most of them were given to me by some people."

"Given to you?"

"Yes. Let me explain. There were some houses around here where women received men for cash. If you understand me."

"Oh... Ok. I understand."

"Some of these women got pregnant and... well, some of them couldn't take care of their kids. So, they gave them to me and my wife. After I accepted the first, my *good father fame* spread through the streets, and other women came and left their kids with me. Usually, they left them with me to look for adventures in big cities far away from here."

"Wow... these kids were very lucky. But didn't you have problems? You know, the kids could suspect or realize they weren't your kids. Did they never ask who their true parents?"

"Hehehe... They knew perfectly well where they came from because I explained it to them since they were toddlers."

"That's insane," Antony was impressed. "Didn't they feel sad? To have been abandoned."

"They got a life. Life is not sad but a blessing. It doesn't matter where these lives came from. They deserved all the truth and happiness I could give them. They all grew up strong and kind. Then they heard that the big cities far from here needed labor. All kinds of good opportunities were waiting for my kids there. And they left, like most people here did."

"And you? Couldn't you go with them?"

"No, no. They insisted, but no. My life is here, it doesn't matter if here is good or bad. I've already fulfilled my role hehehe, educating all my kids. Thirty years ago, my Mary passed away and up to now... I'm here."

"Wow. This is an amazing story, George. Thanks for sharing it with me."

"It's just a story. Hey, look. Another big one. Hehehe."

George caught one more catfish.

After Angela and Nivedita, George was the most captivating person Antony had ever met in his life. Antony's imagination insisted on comparing the way his rich and educated family treated him to how this simple man had raised his sons and daughters. *They are probably happier than me*, was what Antony was thinking, and that made him even sadder.

A rain cloud approached from behind the two men, but neither of them noticed it because the sky in front of them was clear, despite the dark spot in the end of the horizon. The start of the apocalypse was coming, but one of them didn't care, and the other just wanted to fish.

Antony left his fishing rod by his side and contemplated the beautiful landscape for the last time. Even knowing he'd just had one of the best conversations of his life, nothing seemed to relieve his deep sadness. His suicidal thoughts vanished for a while, and what remained within him was a wish to see the world being destroyed by the ego of its few leaders and the naivety of those who followed them.

Suddenly, it started to rain on the lake, drenching those who were on the pier. Antony looked up at the sky and noticed that it was just a passing cloud, but it was enough to soak them through. The rain reminded Antony of his days and nights in the gray Brussels — always drunk and going to all kinds of pubs and brothels. Someone did not believe in any of the directions given by all nations and knew very well that he was an important part of it. One day, a Black Hole appeared in the sky and gave him a chance to save the world. But instead, it just showed that the hypocrisy of his family was even more rotten than he had thought, and his existence was disposable. Nothing else about his family seemed good to remember. Not even his mother who, hours before, had been an example of goodness. Not even Allan Sims who had passed away when Antony was a newborn and left only hundreds of business speeches in videos as his last memories. When he was talking to George, there was a moment when Antony wished that this simple man was his grandfather, but the fear of another disappointment erased this hope.

Antony slowly got up and noticed that George kept fishing as if the weather hadn't changed.

"I... I'm going to my airplane. Maybe take a rest. Are you going to stay here?"

"Yes, I'll stay a little longer. Today has been an amazing day for fishing hehehe... you brought me luck, boy."

Antony felt that it was the last time he would ever see this good old man before the end of the world. His sadness hurt his heart even more, and it was difficult to find the next words to say to his new friend.

"This rain... is this rain not dangerous for you? You can catch a cold if you stay here."

"This rain comes from God, and everything from God is good. I like it. It brings me good memories... hehehe. Just watch the rain carefully and good memories will come."

"I wish I had good ones... Well, see you later."

"See you later, son."

Antony left the pier and walked toward his jet.

His long curly hair and nylon coat were already completely wet as the raindrops mixed with his tears.

This world deserves to end because this world doesn't deserve the goodness of this man, that's what Antony was thinking as he went back to his jet.

CHAPTER 60

Intense flames warmed up the huge war zone where the buzzing of shots and missiles already sounded normal to every soldier.

Among explosions, cries, shots, and soot, a giant robot with a damaged leg came staggering to lean on a big pile of wreckage and sand. This giant robot seemed exhausted, which was just a reflection of the physical conditions of its pilot, Glenda Jackson.

Some mechanics wearing army uniforms went over to try to fix the damage the robot had received as other soldiers surrounded her, giving her cover, protecting their best warrior.

Glenda was panting, and her face was shiny with sweat. Her trembling hand typed quickly on a computer by her side while with her other hand she rearranged her headset.

Two mechanics were trying to fix the giant robot's knee while another climbed up to its waist gears to check on it and fix any problem this big machine could have. This last one looked very nervous, trembling a lot while doing his task.

"General Glenda, I need one hour to get these gears back to their best."

"You look very young, soldier…" Glenda breathed hard.

"Kevin. Soldier Kevin, General Glenda. Special mechanic of the American army."

"How old, soldier?"

"Seventeen years old, General. This is my first mission."

"Gosh… So, Kevin… I don't need them at their best. I need it to work just enough for me to kick some monsters' ass. Just do that. Please."

"Yes, General Glenda."

The youth of the soldier impressed Glenda because he reminded her of the guys from the community who would spend all day playing soccer and having fun on the streets. *He doesn't deserve to be in this situation, actually nobody does*. That's what Glenda thought as she watched the boy fix the giant robot and tried to contact the air force.

"Air force, Air force... Can anybody hear me?"

"Eagle 8 here. Go on, General Glenda."

"Prediction to arrive at the destination, Eagle 8? The situation is not very good here."

The noise of an explosion a few yards away from Glenda rang, and the soldiers around her started shooting their machine guns.

"Approximately twenty minutes, General Glenda."

"Twenty minutes! I'll probably be dead before that."

"We are going as fast as we can, General. Just stay strong in your position."

"'Just stay strong!' I doubt you would say that if the President of the United States was by your side," Glenda sounded sarcastic.

"General Glenda, I guarantee that the President is aware of the situation in Cloud Base and —"

"Ok, see you later... Damn it."

Glenda hung up abruptly and punched the metal structure of her cabin furiously. But things seemed to only get worse.

One of the four hundred-foot giant Angels, who were nicknamed *Big Monsters* by the soldiers, finally moved and flew toward the building — and stood at the top of Cloud Base. It was the golden one, wearing a sphinx head. This Big Monster slowly struck its spear with the screw-like tip onto the roof of the building. There was a hand crank in the middle of the spear that the Big Monster slowly turned. This way, he made the tip spin and drill on the thick roof that was made out of metal and concrete.

Glenda looked desperate when she saw this scene and turned her headset back on.

"Bros, can you see that? The Big Monster on the roof. All drones and tanks close to the target, go and shoot down that motherfucker."

The drones obeyed the General's orders immediately; they flew directly toward the giant and shot at him with no mercy.

Glenda watched the violent attack of the drones with hopes that everything could be gotten under control. Several explosions covered the Big Monster, and Cloud Base's roof seemed to finally be protected. However, none of the bullets and missiles affected the giant warrior who didn't stop his terrible mission — despite the huge fire that covered his whole body. In that moment, this Angel looked like a gigantic statue spreading flames and smoke with the movements of his large wings while several drones flew over him like a swarm of wild bees.

The other one, the silver and red Big Monster who had an eagle helmet, stayed in the middle of the war zone and started striking tanks, soldiers, and drones around him like a crazy hockey player hitting army toys.

Behind the pile of wreckage and sand, Glenda looked frustrated and without hope, watching how the Big Monsters seemed indestructible and how the troops were vulnerable.

The soldiers on the robot's leg and waist worked quickly against time and the death all around them.

Soldier Kevin didn't stop trembling while tightening some large screws. But this feeling didn't keep him from starting a conversation with his General just to try to calm himself down.

"General Glenda, can I ask a stupid question?"

"Everything in this place is stupid. Feel free to ask."

"What are we doing in this war?"

Glenda took a deep breath, knowing that wasn't a stupid question, just hard to answer.

"Bro, if I knew this answer, I wouldn't be here. Now, focus on your work."

"Yes, General Glenda."

She made the most of the few minutes she had before starting to fight again to take a breath. She whispered to herself:

"Oh Lord, Antony would have already fixed this robot using a paper clip. Where could this little chicken be?"

CHAPTER 61

Despite the light rain falling like water from a shower, it was enough to create several puddles along the dirt path Antony walked through. He stepped on them, not worrying about getting his white sneakers dirty. In fact, Antony worried about nothing as he walked, crestfallen and soaked through, toward his jet.

It was a bright day, and a soft breeze blew through the pines that formed a corridor around the wet path.

"Just watch the rain carefully and good memories will come."

Antony whispered to himself what George had told him, knowing that was the last thing he'd hear in his life.

Instinctively, Antony looked up at the cloud that was pouring water on the forest and tried to think of some good memory of his family. But he knew that any good memory now would seem fake — despite the good intentions, except from his grandfather.

Allan Sims had died when Antony was a newborn, leaving an empire for his family to administer, but not the image of a loving granddad like the ones who attended Christmas parties carrying lots of presents for their grandchildren. No, he wasn't like that. Antony had only seen his grandfather through hundreds of business speeches, interviews, and documentaries about his professional life. All of these showed him as a powerful rich man who was incredibly eloquent, able to answer any question about the business world, leaving all reporters and opponents not knowing how to react.

However, for Antony, his grandfather was just a character he saw on videos on the internet and television. A man who looked like a big iceberg, cold, calculating, and ruthless to reach his megalomanic goals. A man who had lived for more than one hundred years thanks to the miracles of medicine that only he could pay for.

The heir of Sims Corp couldn't think of a single good memory of his grandfather, because, for him, they didn't exist in reality.

"Just watch the rain carefully and good memories will come."

This quote insisted on buzzing around in Antony's mind until he arrived at the old abandoned hangar where his blue jet was landed.

The door of the vehicle was open, and Antony got in without worrying if he would drag water inside.

Inside the jet's cabin, he saw the three moggie cats sleeping on the floor, close to the cardboard box where they had been caged for days by Uncle Peter.

Antony sat down in his habitual armchair like a depressed patient on a lounge chair in a psychiatric office. He relaxed his wet head on the armchair backrest and watched the raindrops fall outside the window. The sounds of the raindrops falling created a kind of melancholic symphony that would make anyone sleepy, as if each drop was a piece of his good memories. Unreachable memories.

"Just watch the rain carefully and good memories will come," Antony's mind kept spinning this sentence around for a few minutes.

"Just watch the rain carefully and good memories will come," his voice sounded hopeless.

Antony's eyelids were heavy. Exhaustion was taking over his body little by little, until...

"Just watch... memories?"

Suddenly, Antony opened his eyes wide, as if he had just been startled awake from a deep sleep. He had an epiphany — an idea that seemed totally crazy but that pushed him out of his depressed state.

"Memories! The symbol? Yes, the symbol. I remember."

Antony took his tablet that was next to the armchair and started typing quickly, looking for the files that had Angela's documents and medical reports. The files his cousin Christine had sent him to kill his self-esteem and destroy the loving relationship he had with his mother.

Then, Antony opened the first medical report and looked carefully at the symbol stamped on the bottom of the document.

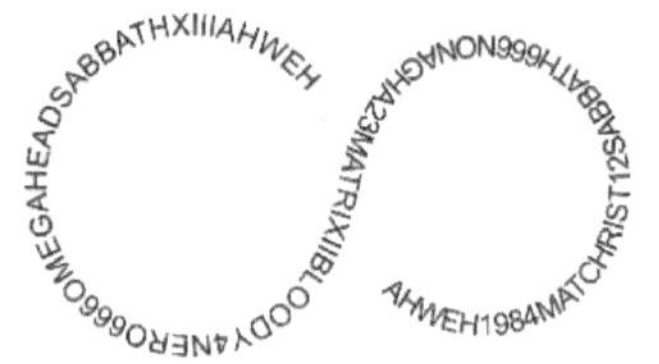

"It's the first time I've seen the Corporation's symbol this way, made out of different numbers and letters... and in a medical report? That's insane, but I'll try it."

His depression vanished for a moment, and his typical curiosity came back in full force.

With an image editing program on his tablet, he scanned the symbol and took it off the medical report. With his fingers, Antony could manipulate and distort the symbol any way he wanted, as if it were a piece of wire on the screen. Then he stretched the symbol and formed a line.

AHWEH1984MATCHRIST12SABBATH666NONAGHA23MATRIXIIBLOODY4NERO666OMEGAHEADSABBATHXIIIAHWE

"I'm right. This is… it looks like the link to a video."

When someone uploaded a video to the internet, the website that hosted the video automatically created a link with random numbers and letters. Then, anyone could copy this link and send it through an email, text, or post it to some social network. But a distorted link stamped on a medical report looked odd, especially when it formed the symbol of the most powerful corporation in the world — and Antony noticed it.

"Well, there is just one way to discover if it's true."

Antony copied and pasted the link into a search engine. Not Goinggo.com, the only search engine that, of course, belonged to the Sims family, but to Horus.net, a search engine created by Antony and other programmers from Sims Corp that was able to find information beyond Goinggo.com — especially inside closed systems of companies and deep web around the world.

If the video wasn't hosted in a computer inside Cloud Base, Antony could find it.

"Ok, it seems crazy, but I have to try."

Antony pressed the search button, and after a few seconds, the search engine found a video named 0.1 in a Sims Corp's office in Jerusalem.

"Damn. I found it. In an office in Israel? It was always the place the family forgot about. Nobody liked to work there. Just... just Allan."

For a while, the video was frozen, with zero views and ready for someone to play it for the first time.

The frozen image on the screen made Antony nervous. It was a self-recording made by Allan Sims with a phone in a place that looked like a hospital room. His grandfather looked sad, and behind him, there was a woman in the back trying to sing a baby to sleep.

"Wow, what is this? I don't remember a video where my grandfather looked so down. Well... another disappointment from my family won't make a difference to me," Antony took a deep breath. "Ok, here we go."

Antony played the video, and Allan Sims started talking in a strong but calm voice, like he was getting something off his chest. A behavior which he never showed in his business speeches and interviews.

"Hi, Antony. It's your grandfather speaking here. If you're watching me, you were smart enough to notice one of the symbols that I hid in some places on the internet. Actually, very smart. Anyway, you're probably asking yourself 'why did he do this?' The answer is, because I wanted to protect you. Protect you from the evil that has ravaged our family. This is a well-known story that has haunted many families. The patriarch of the family struggled a lot to build a rich patrimony for his family to stay together and keep all this wealth. Then, what does the family do? They start a feud between themselves for more power and a larger percentage of the patrimony, then they disunite. If you found this video, certainly one of the Sims' family members wanted to kill you. Kill your soul... kill your will to live. Never let that happen... Never... Well, how can I say this? There are many things for me to tell you, and you probably still have many doubts about why I created Cloud, doubts about your mother... and father. And I'll explain everything. Or at least, I'm gonna try."

Allan Sims took a deep breath and looked slightly excited.

"I'd better start telling you about myself. Well, I was born in South Africa, specifically in the capital, Pretoria. My parents were American diplomats, and we lived a perfect life, especially when they discovered... how can I say? My gifts. 'Their only son was born super gifted, and he'd probably have a brilliant career,' was what everybody around my family said. Everything was going amazing when my loving mother and father passed away in a plane crash when I was a teenager. Then an excruciating pain took over my heart. Of course, you don't know this story because I made sure to delete any traces of it from the internet. Anyway, after I finished my graduation in San Francisco, I had an idea... for me, the best idea in the universe. A way for no one to ever feel the same pain I felt when my parents died... My parents... good people, loving, religious, honest. 'Why did they die? Where was the Lord when they were dying?' I thought about it every single day. Furthermore, I decided to dedicate my life to creating something that would allow humanity to overcome death. A machine that could take away our fear of losing the ones we love, take away our fear of death. Cloud... that was my idea, a system that worked with a kind of algorithm able to keep the human consciousness and memories inside a virtual world. That was an amazing creation. There no longer had to be a choice between living and dying. There was just life. People and politicians around world were very excitedly supporting it. Finally, I built my family. I know, I took too long because of all the years of hard work and stress, but finally, I built a family, a family that I missed so much. Everything seemed wonderful, and my creation was changing the world... but for the worse."

Allan's voice lost its enthusiasm.

"When I realized, I had lost control of my creation and family too. It was too late. Cloud gave politicians and other powerful businessmen a way to control society without violence. People got pleasure out of keeping themselves in stupid arguments about prejudice. Religions had vanished, and supposed good leaders were being accused of being dictators by the media... a media controlled by the same politicians and businessmen who enjoyed the resources of Cloud. You're probably asking: 'Why did he let these kinds of people control the world?' Well... I needed help to build Cloud, and I believed that a world free of the fear of death would cause a commotion in their hearts. I always believed in the goodness of our leaders... I was so stupid. Years passed, and the situation had become bad. I thought: 'It's a matter of time. Humanity will rewrite its own history.' But no... Furthermore, I have been dying slowly, and I don't want my mind to end up inside... how can I say? Inside my own creation... inside Cloud. Because this place... this place became a mechanism for powerful people to control society, a control they took in exchange for hope, the hope to live in a perfect place. But it's not a perfect place. It's... millions of tangled cables, circuits, and algorithms creating an illusion. It's not a life. I thought every single day about destroying it, but I couldn't. Because Cloud became so big, my wealth became so big, and I couldn't ruin my family and other families who depended on my creation. I love my family even knowing they are evil... even knowing I'm not a loving father..."

Allan's eyes started to fill with tears.

"How can I say? I was losing my family, and I couldn't build a new one. And as evil as my family is, I was unable to stop loving them despite not trusting them. I'm dying, and I'll leave behind a totally rotten world. But suddenly, I had an idea, an amazing idea. I needed a new heir, someone good, someone intelligent enough to save the world from the shit that I created. I know that sounds unfair, putting this responsibility on someone else, but it was my last hope. I needed an heir and the only person who could give it to me was Angela. Antony... if you're watching me now, you probably know that it was hard to convince your mother to have you, but I convinced her. Her body was destroyed; it wasn't ready to hold a pregnancy. Your mother had a lot of health problems due to... to an unhealthy lifestyle. But after much effort, me and my doctors found a way, a way for you to exist. Through a scientific analysis... In a genetic point of view, you're my grandson... and my son too. I know it sounds insane, but I was desperately needing a good person to save the world and our family. And you're probably asking: 'How does this old man know if I'm a good person?' Well... look at that."

Allan turned his phone a little towards the woman who was standing, rocking a baby to sleep. The woman was Angela, and the baby was Antony. Angela wore a white hospital gown, and Antony was covered in a white towel. Angela's face shone with happiness, and little Antony's big smile seemed to last forever in this beautiful scene which Allan Sims had always dreamed of witnessing since his parents passed away.

Angela lifted little Antony and let the emotion from her heart express a loving language which only a mother and her baby could understand.

"Hey, are you my baby? Yes, you're my baby. Mother's little baby. I know you love me, and I love you too, you know —"

Angela hugged Antony as if she had been ready for this moment her whole life, and the baby rewarded her with a big, happy smile.

Then Allan pointed the phone camera back to his face, his eyes tearing up.

"Did you see that, Antony? It's... it's you and your mother. You're a good person because you transformed your mother into a good person. I can see it in her eyes. She has changed, changed for the better... thanks to you. Well, how can I say? I only have one request for you... Be a good person, my grandson. Be better than me. If you are good, you will know exactly what must be done... I love you, Antony. I love you so much... wherever you are."

Allan turned off the recording, and the screen of the tablet in Antony's hands became totally black.

The rain outside continued its melancholic symphony, falling over the jet's fuselage, and the black screen of the tablet was getting wet with a lot of drops — drops of tears.

CHAPTER 62

After a failed request for mercy to her master, Muriel crossed the Black Hole and went back to the war zone in the Nebraska desert. Her flight looked like a dance over explosions, shots, screams, and death. While she watched the huge, horrible scenery of destruction and fire below, a terrible prediction came to her mind. She was sure that this hell below was just the beginning, and shortly this scenery would grow more and more until it took over the whole planet.

The red gleam from the war lit up her desperate face, and in a low whisper, she expressed her agonizing doubt.

"My Messiah, where are you now?"

CHAPTER 63

The rain stopped, and there were still many puddles on the wet path, but George walked through it with no worries, feeling very satisfied with himself after a great morning.

His bucket was full of fish, and he'd met a new friend. The day couldn't be better for this kind old man — and there was one more surprise.

As he walked, carrying the bucket in one hand and the fishing rods and tackle box in the other, George saw a cardboard box in the middle of the path. So, he decided slow down, until he noticed that there was something written on the box in red ink.

George was suspicious but didn't lose his typical smile.

Suddenly, a moggie cat got out of the box, leaving George mesmerized. He dropped his fishing gear on the ground and walked very excitedly over to another new visitor, like a child receiving a present.

"Hey, friend. How are you? You are so cute, did you know that? Hehehe."

George crouched down and took the cat in his arms. Then he noticed there were two more meowing inside the box — which made him even more excited.

"Oh... two more friends. You're certainly a gift from the Lord. Let me see what is written here."

On the cardboard box was the following message:

They are nice friends. Care for them like you cared for your kids, please. P.S: Tony Sims.

The old man got emotional and hugged his new friends tightly, who returned his affection caressing his face with their heads.

Suddenly, the buzz of a powerful engine sounded through the air, catching George's attention. He looked up and saw a blue jet flying fast toward the black dot at the end of the horizon.

The old man, filled with happiness, followed his friend moving far away with his wet eyes and said:

"Good boy!"

CHAPTER 64

A convertible army pickup raced among mountains of wreckage through winding curves until it reached a path where there was a giant Angel wearing an opaque golden armor with a helmet that looked like a gorilla's head, showing its teeth. Also, the giant was holding a metal morning star, and its large gray wings were open, signaling that he was ready to attack.

In the distance, explosions, shots, the buzzing of drones, and screams were like the soundtrack to this horrible scene where soldiers and giant-winged creatures faced each other.

The pickup stopped, and the driver said to the three soldiers behind him, ready to shoot:

"Hey guys, that's the big one."

"We see it, and we're gonna attack with heavy artillery," one of them replied, "drive fast towards this damn monster."

"Ok... GO!"

While the pickup raced, accelerating more and more, the giant Angel, wielding his morning star, ran against the vehicle. Two of the soldiers in the back seat readied their bazookas, and the other one readied a grenade launcher.

The driver, with bloodshot eyes, kept his focus on the dirt and stone path until he shouted out the order:

"FINISH THIS MONSTER, GUYS!"

The three soldiers shot at the Angel turning him into a big wall of fire and smoke — a barrier that closed the only path completely.

Taken over by a survival instinct, the driver executed an emergency maneuver.

"HANG ON, GUYS," the driver shouted out.

He hit the brakes abruptly in front of the big wall of smoke. It was so intense that the Angel seemed to vanish behind it.

The four soldiers breathed nervously and looked up, trying to identify some trace of their opponent.

"We got him?"

"I hope so."

Like a comet coming down from space, the Angel's morning star appeared out of the wall of smoke and hit the side of the pickup hard. The vehicle spun, throwing the soldiers to the ground.

Panic took over the soldiers. One of them had his leg stuck under the wreckage of the vehicle, and the others were seriously injured. Two of them were still holding their pistols toward the Angel. They were trembling, covered in dust, sweat, and blood, knowing that was a useless attitude.

The giant Angel appeared, staggering and injured as well, covered in flames, but strong enough to wield his weapon.

He lifted the morning star again, ready to slash at the vulnerable soldiers, and for a moment, they were sure they were going to close their eyes and only open them again inside Cloud Base Two. But suddenly, a long sword blade came quickly out of the smoke and ripped the Angel's arm that was holding his weapon. The hand still held the morning star tight as it fell heavily to the ground.

The giant winged warrior staggered away from the smoke in a useless attempt to save himself when the same long sword blade struck him in the middle of his chest. Viscous blood gushed from his chest, and the giant screamed, a sound that was like an eagle's cry. Then, the giant robot responsible for this feat came out of the smoke and took its sword from the Angel's chest — that fell lifeless to the ground.

The robot turned, and its pilot asked the soldiers below:

"Are you ok, bros?"

"For now, yes, General Glenda," one of them answered.

Glenda's giant robot was a little damaged but ready for combat, unlike its pilot, who was totally exhausted.

"Ok, bros... now we must go back to —"

"General, look at that!"

One of the soldiers pointed up, causing Glenda to look up in the same direction as well.

About twenty Angels were positioned, standing in line in the air like wild birds, moving their wings intensely to maintain stability and pointing their spears and arrows directly at the humans below.

Winged women and men, all muscular with different skin tones and various armors. Some wore freakish helmets in the shape of different beasts, while others revealed their brave faces and bloodshot eyes.

Like magic, the tips of their spears and arrows ignited in flames, assuring Glenda and the other soldiers that they were ready to attack.

Glenda wielded her sword, still soaked in the last opponent's blood, and positioned herself between the imminent attack and the desperate soldiers. One of them pointed his pistol toward the winged warriors, while the other two tried to release the one who was stuck under the pickup.

The moment was tense, the Angels merely waiting for the best moment to launch their flaming weapons. They were close to their targets but not enough for Glenda's robot to reach with its sword. Despite the robot being armed with missiles and machine guns, Glenda knew she couldn't hit everyone at once.

The Angels looked like they were armed with torches, and Glenda's face sweated more and more. She didn't blink her eyes and kept her trembling hands on the robot's controls, just hoping that her instincts guided her to act at the right moment.

"Come on, ugly Angels. End this damn game."

Tension, nervousness, and a warm wind brought up the smell of blood, while the two soldiers tried vainly to lift the upturned pickup to save their injured friend.

"COME ON, MONSTERS," Glenda shouted out loud when she was surprised by surprise by an odd and unexpected buzz that became increasingly loud.

"What the hell is —"

A miracle? An Angel? A big blue stain in the air! An aircraft crossed over the sky at high speed and ran over the lined-up Angels, throwing them far away and driving away the others — like it was big bowling ball and the Angels mere pins. Only some feathers floating in the air and some weapons fallen on the ground remained.

Glenda still didn't know if it was a miracle or a lucky strike, but she felt like she'd already seen this aircraft.

The soldiers below managed to save the one under the wreckage of the pickup and looked up at the sky, full of doubts.

"What just happened, General?"

"I don't know, but I have a bad feeling. Ok, bros. You're gonna find a rescue team... two minutes west of here. I have to check on something. GO!"

Glenda looked desperate and stopped paying attention to the men she had been protecting as she piloted her robot, running up to the top of the next pile of wreckage. There, she stood up, trying to locate the jet that had probably just saved her life and the life of the other soldiers.

"Oh, no... it's impossible."

She saw a dark blue Bombardier Global 1K with white stripes around it and a white infinity symbol on its fin. This modern machine stood out in a dark scenery where giant Angels and robots fought among flames. It flew above it, dodging the drones and Angels like a fighter jet. Then this aircraft went up and flew over Cloud, close to the sphinx-headed Big Monster who was drilling on the roof. This giant Angel noticed it flying around him and tried to hit it with his spear. But the jet was quick, and the Big Monster was slow.

This aircraft was trying to land on top of the building, although the Big Monsters and many Angels around it wouldn't let it.

With sword strikes, burning arrows, and spears, the Angels tried vainly to shoot down what seemed to be a new enemy to them. The jet was fast, resistant, and able to dodge quickly thanks to its powerful retractable turbines.

"What are you trying to do?" Glenda whispered as she stayed on top of the wreckage. She was scared and nervous, watching from far away as the jet did doing daring maneuvers to escape the agile armed Angels up in the sky.

After many attempts, one of the Angels, holding a big ax, hit one of the jet's wings. The aircraft lost control and fell down spinning — away from Cloud Base. Due to its retracted turbines, it gained a little stability as it descended until it hit ground violently.

Inside her robot, Glenda was frozen, just watching the jet spin and skid on the asphalt, spreading sand and dirt around it. The fuselage under it was destroyed, and flames rose from the turbines.

"Nooo... gosh. You can't be in there," Glenda whispered.

The dark blue jet finally hit a pile of wreckage and stopped, covered in dirt and smoke from the turbines that were in flames. However, the worst was yet to come. The jet was close to the silver and red Big Monster, who had a medieval helmet that looked like an eagle's head, and this Angel was walking towards the jet.

The soldiers around the crash site stayed in their positions in the war but were confused about the aircraft that had just fallen.

With slow steps that made the ground quake, the Big Monster approached, holding his long and heavy spear ready to smash the fallen jet.

Glenda was about three miles away from the fallen aircraft and unable to do anything quickly, except warning the soldiers that were close to the crash site to move away from it. Her trembling hand turned on the radio signal, and her nervous voice shouted out a desperate order.

"Everybody near the fallen jet, get away from there. RIGHT NOW. This —"

Several soldiers ran away quickly, but it was too late.

The heavy distorted tip of the spear fell slowly but terrifyingly toward the jet, that was already in flames and covered with smoke.

Glenda was so nervous that she didn't notice the jet's side door open slowly — just the spear directly hitting the flaming turbine.

"This damn machine is NUCLEAR... get out of there!

An inevitable blow hit the jet, and an immense explosion rose like a sun in the middle of the war zone. A huge ball of fire blinded Glenda and scared even the soldiers that were far away from there. The Big Monster was in flames, and the explosion created a kind of huge circle of fire that spread out from the crash site.

A warm wind blew through the horrible scenery, and Glenda's tears mixed with the sweat on her face. Her eyes lost any trace of hope.

"Oh gosh. That can't be happening. It's worse than a nightmare. AHH... what's this... Oh, my ears!"

Suddenly, a sharp buzz came from her headset, irritating her exhausted mind. Then the buzz stopped, and Glenda could hear a confident voice calling.

"Glenda, can you see me?"

Glenda's reason didn't want to believe the voice from the headset, but her instincts recognized it immediately.

"ANTONY... Is that you?"

Glenda's eyes focused on the circle of fire until she noticed something coming from the flames. At first sight, it looked like a rocket on two wheels being ridden by a thin guy wearing a red helmet. But Glenda quickly realized who was escaping from the flames.

Antony Sims was riding his dark blue Kawasaki Ninja H5R that seemed to fly from the fire and land in high speed on ground. He crossed the piles of wreckage and the fights between humans and Angels fast, skillfully dodging any obstacles on his way to Cloud Base.

It was still hard for Glenda to believe who had shown up suddenly in this horrible scenery, but she replied to the voice in her headset.

"Yes... Yes, I can see you, Antony."

"So, move your fat ass and come cover me right now."

"Oh gosh, it's fucking unbelievable."

Glenda's giant robot went down the wreckage, running as fast as possible toward where she caught sight of Antony riding his motorcycle. She dodged quickly past any giant Angels in front of her, jumping over wreckage and destroyed army machines.

Through all the obstacles of this huge war zone, Antony Sims looked like he had recovered his liveliness, totally motivated and full of energy. His speech sounded like his typical sarcasm, and his mind worked fast, doing many calculations. The visor of his helmet, connected to the on-board motorcycle's computer, showed all the information he needed about what was around and above Cloud Base. He managed to hack a Sims Corp satellite before he left the jet, using it to get his bearings in the middle of this hell and keep in touch with Glenda.

The wheels of the Kawasaki Ninja H5R ripped through the ground, leaving behind a trace of smoke and debris.

Some soldiers called through their radios on the open signal, asking each other questions like, *Who is this madman riding a motorcycle?* or *Where did this crazy guy come from?*

In highs peed, Antony found the right way and, for a moment, raced free until he saw a dangerous obstacle before him. Three low-flying

Angels, armed with swords and spears, were coming toward him fast, ready to attack. Only a few seconds separated them from an imminent collision when Antony noticed just ahead, on the ground, full soldiers' corpses and Angels shot down, a machine gun. He thought quickly and leaned his body without slowing down his motorcycle, catching the machine gun. With confident movements and a lot of agility, Antony held the weapon with one hand while the other kept controlling the motorcycle.

"Come on, crazy Angels. Say Hello to my little friend!"

Antony shot mercilessly at the three low-flying Angels coming in his direction. The shots hit the winged warriors hard, and they couldn't maintain their trajectory against the bullets. Scared and injured, they flew away, leaving the path clear for Antony.

Glenda ran panting, without stopping, but still far away from her friend. She kept in touch with him the whole time, especially when she heard shots coming from her headset.

"Antony… what the hell are you doing?"

"Just obeying the last order my grandfather gave me. Being a good person. Listen to me, we can end this war, but I have to enter Cloud Base. And to get in, I need your help. Go straight to road number eight, about two miles west of where you are now. I'm gonna meet you there."

"So… you have a plan. Right?"

"Yes, but you're the General. You must ask your soldiers to keep the road free of Angels. That road is our best option to get into the building. Moreover… keep some missiles; we'll need them."

"Ok, bro."

Glenda didn't have any other option except to believe her instincts, which told her to trust Antony. She was breathing fast, and her heart beat to the rhythm of the machine guns' shots that rang through the air. She opened the communication signal of her headset so that every soldier could hear her.

"Attention everyone, please. Attention. Change of plans. Everybody positioned around road eight. Get this road free of Angels. Me and the guy on the blue motorcycle need this road free. Again, get road eight free of Angels."

One of the soldiers contested his General, and others replied with the same doubt.

"Why do we need to protect this guy, General?"

"He's... he's Antony Sims, and he can save this damned day."

The name of the Sims Corp's heir sounded like a blow of hope to every soldier who had almost given up fighting against these winged giant Angels. *If the youngest son of the Sims Family is in this damn hell, there is still salvation*, was the idea that came to the soldiers' minds at this moment.

Another soldier's voice shouted through the communication system.

"You heard our General. Protect the road and let the guy save the day!"

A feeling of victory pushed all soldiers immediately toward the surroundings of road number eight in their army vehicles. Giant robots hurried to keep these new positions along the borders of the road, and army drones were already flying over it.

Antony sped up increasingly, facing all kinds of obstacles he had never imagined, even in his craziest dreams. He dodged a giant Angel on the ground who tried to hit him with a long club, then he escaped flaming arrows shot by another Angel, and a drone exploded this winged warrior with a missile.

"Wow, I feel like I'm inside a video game, but at the same time, I'm sure it's real life. Ok, here we go!"

Seeing all this destruction made Antony's mind remember how Allan Sims dreamed of a better world and how the other members of the family, along with rotten individuals like Yelena, had transformed his dream into an immense piece of shit. However, something that George had said insisted on buzzing around his mind too. *We're just where we have to be, doesn't matter if our actions were good or bad. These actions always lead us to do the right things in the right place.*

"If I deserve to be here, I'm gonna do the only thing that can save this world."

While Antony crossed all kinds of obstacles, he could hear some soldiers shouting things like *Go Tony* and *Save our day*.

He saw the edge of road eight, a large freeway that looked like a dry artificial river made out of concrete, and noticed a big piece of wreckage that formed a kind of ramp.

The machine gun was still hanging by a sling on his side, and his heart beat fast, but his adrenaline was so high that he hadn't noticed it yet.

"It's gonna be insane and dangerous too."

Antony sped up, raced between two giants fighting and a tank shooting. Then, he was on the ramp. Adrenaline, tension, and danger... he seemed to fly for a long hour, but it actually lasted only a few seconds until he hit the dirty path hard, maintaining high speed on the road. The way was clear, and at the end, Cloud Base seemed like a faraway place, impossible to reach.

Despite the road being free, just ahead on both sides and in the sky, many fights were happening, creating a kind of long path surrounded by war. The order from Glenda was being fulfilled.

Antony stayed focused on his way until he saw the giant Angel ahead. This warrior was wearing a graphite-colored armor, and his helmet, resembling a bull's skull, was as white as his wide wings. This winged giant looked directly at the motorcycle approaching him and was getting ready to attack with a hammer that had a long handle. Antony looked up to his side where Angels and humans were fighting and got an idea. He held the machine gun again and started shooting right at the head and chest of the Angel.

"You're damn ugly, my friend. Let me fix this horrible face," Antony didn't show any fear.

The winged Angel lifted his big hammer, ready to crush his target who was approaching and shooting with no mercy.

A sound like a scream from a savage eagle echoed through the air, and the Angel struck his target violently with his hammer, but Antony was even faster. The motorcycle escaped the hit and raced between the legs of the winged warrior like a rocket. Then Antony turned the motorcycle abruptly and skidded on the ground, shooting at the giant's face nonstop, which didn't hurt him very much.

The giant Angel turned furiously, looking straight at Antony and gave another terrible scream. He was ready to run toward his target and try another blow.

Antony stayed on his motorcycle, still shooting at the Angel who got angrier by the second, but the bullets finally ended, and Antony threw the machine gun on the ground. Even then, Antony kept facing the giant.

"Yes, my damn friend. Just keep looking at me and say goodbye."

The giant Angel was getting ready to run toward Antony, but he didn't notice Glenda piloting her giant robot, jumping right at his back. She came spinning like a ballet dancer but wielding her sword. The long blade hit the side of the Angel's head hard. Then the giant fell, lifeless, leaving a trail of viscous blood in the air.

The giant robot stood up, looking exhausted, which was just a reflection of the physical conditions of its pilot.

"Great strike, Glenda. Like that first time. Remember?"

It was hard for Glenda to believe Antony's calm sarcasm in this piece of hell in the middle of the desert. She sounded angry but happy at seeing her friend again.

"You're fucking crazy, little chicken. Where were you all this time?"

"Hmmm... learning how to fish. Anyway, keep running behind me. I'm gonna explain my plan on the way."

Antony turned his blue motorcycle and sped up toward Cloud Base.

"Bro... I hate you. But I love you at the same time."

Then Glenda ran fast right behind him while, up in the sky, a winged warrior watched carefully the actions of both on the road. Muriel was filled with emotion when she realized her chosen one was in action again.

"My Messiah, you came back. My God heard my prayers."

The battles got more and more violent around the road as Antony raced towards Cloud Base with Glenda running behind him. Shots, explosions, flames, and a hot wind covered all their trajectory.

The hearts of Glenda and Antony beat fast, like the engines of their machines, as the adrenaline kept them focused on the road. All the information about Glenda's robot, the distance of the road, and the security system of Cloud Base were displayed to Antony on the green visor of his helmet, while his mind calculated his next actions and all possibilities. Any mistake or some Angel who blocked their way could ruin the plan to get into the building.

"Ok Glenda, it's now or never. Listen to me. This road takes us to one of Cloud Base's massive locked gates. When we get close to it, you have to keep your target on it. You have two more missiles on your shoulder, and that's enough to open it. Just wait for my command to shoot."

"Ok, when we get there, I stop and shoot —"

"No, no, no… this gate leads to a long corridor. When you destroy this gate, almost at the same time, another bigger one… a second massive gate will close right behind it, like an immense guillotine. Let me sum up my plan. You explode the gate; I get in and race through the corridor before the second gate smashes me. That's it."

"THAT'S IT? What if you don't get in?"

"But I will. So, get your target ready. You must be close to be sure to destroy the gate completely, and I must be fast enough to cross the corridor in time. When I say *fire,* shoot the missiles."

"Oh gosh… let's go, bro."

Antony sped up more and more, and Glenda kept her robot close to him like a frenetic runner chasing a toy motorcycle.

Speed, tension, adrenaline, fear, and hope. Many feelings were mixed in this dramatic moment.

Glenda set a visor from her headset on her right eye and aimed the gate that was just ahead.

"It's ready. Just waiting for your signal."

"Ok."

Speed, tension, adrenaline, fear, and hope. There wasn't any more reasoning now. Just their instincts drove them.

"Wait... wait. One... two... three... FIRE!"

Glenda shot two missiles that flew right by Antony's head.

Right in front of his eyes, there was a big explosion that made Antony speed up to the maximum. His motorcycle seemed to fly at this moment as he raced into the flames.

When Glenda saw her friend disappear into the fire, when she realized that her robot was too fast. The crash against the fiery gate was inevitable. She stopped the metal legs abruptly, making the robot roll over several times until it crashed with its back on the gate.

The crash damaged the robot completely, and little by little, it was being covered by the flames from the gate as Glenda was unconscious, stuck inside it.

She didn't know if Antony's plan had worked or not, and her body was so exhausted that she felt unable to save herself. Her face was covered in sweat and blood, her eyes were slowly closing with exhaustion, and even the noise of the explosions and shots couldn't wake up the best warrior in this infernal war from her exhausted state.

***

The corridor was long and wide, with massive walls and a smooth concrete floor. The lighting system was off, but at the end of the corridor, there was a huge lit hall — that was enough to see the right direction.

An intense explosion happened at the exit gate, and a big ball of fire arose. From its flames, Antony Sims raced in as fast as his motorcycle's engine could go. His nylon coat was in flames, but the extreme speed put them out.

"DAMN... it was more than insane. Now —"

A sharp metal sound echoed through the corridor, and Antony saw a heavy steel gate automatically closing, like an immense guillotine blade.

"GO, GO, GO —"

The motorcycle's engine sounded fierce, and through pure instinct, Antony noticed that the chance to get into the hall before the gate closed was minimal. So, he leaned his machine and lied as close to the ground as he could.

Everything happened in a second. The velocity power pushed Antony, who felt his machine get away from his body. He spun and spun increasingly, skidding on the ground and feeling his body getting hurt and burned.

He kept spinning and skidding out of control until he saw the heavy gate in front of his eyes almost smash him. But the velocity force pushed him in time, and suddenly, Antony just heard the heavy gate hit the ground right behind him. He spun, skidding a little more until he stopped in a huge, white and gray hall that seemed to be an area to transport heavy equipment.

The lighting from this high place dazzled Antony's eyes behind the visor of his helmet. It took a while for him to believe the feat he had just accomplished.

"I got it... wow... it was damn insanity. I got it!"

Antony got up slowly and felt the pain cross his whole body. Parts of his arms, legs, and nylon coat were burned. His fast heartbeat made him feel weak. Also, his motorcycle was fallen by his side, and its engine slowly turned off.

He watched around the huge area and saw many elevators a few yards ahead of him.

"Ok, from now on it'll be easy, I guess," Antony was panting as he whispered. "I need to reach Christine's office to get into Cloud. Inside there I can end this damn apocalypse."

But suddenly, Antony heard several steps fast approaching at his side, then he saw about ten soldiers wearing black uniforms and helmets, pointing their guns at him. One of them shouted:

"FREEZE... hands up, man!"

Overwhelmed with a strong pain that affected his body and a feeling of disappointment as well, Antony just said in a tired voice:

"Well, I forgot... there's nothing easy for me here."

CHAPTER 65

Glenda's eyes slowly opened, and she felt almost lifeless — she was weak, very weak. The cabin of her fallen robot had been destroyed, and the dented hardware kept her from moving, especially her legs. She got stuck in her own machine as flames raised from the gears. And the situation would only get worse.

Heavy steps approached, making the ground shake like an earthquake.

Glenda saw one big foot that looked like an immense eagle claw made out of metal. That foot belonged to the Big Monster, one of the two giant Angels who was four hundred feet tall and wore a medieval helmet that looked like the head of an eagle. Its tall body, wearing a silver and red armor, looked like a building damaged by shots and explosions. But its hands held the heavy spear firmly.

Escaping from the distorted gears was impossible, Glenda knew that, and she was ready for her unbearable pain to end.

"Come on, bro. Come on… I'll hate to wake up in this fucking Cloud, but I can't resist," Glenda had lost all hope.

The Big Monster held up his weapon and looked right at his target lying down by the gate that was still in flames. The inevitable strike was imminent.

Glenda closed her eyes and hoped desperately that she'd wake up in a good place in Cloud or what she believed to be heaven, when she suddenly heard someone landing on her robot cabin. She opened her scared eyes and saw a winged woman wearing a silver armor with shades of blue — also, long black dreadlocks on her head that gave her a brave appearance. She was Muriel.

The immense distorted tip of the spear fell like an asteroid on the pilot's cabin, and only Muriel stayed between Glenda and the terrible weapon.

Muriel looked serious and confident when the spear reached the point where it would hit her, but with incredible strength, she stopped the spear with only one of her hands. The Big Monster forced his weapon down, but Muriel was very firm as the hardware of the cabin slowly crumpled.

Glenda was frozen, mesmerized, just gazing at the power of this winged woman as she felt her body being crushed by the cabin.

Muriel faced the eagle head as she took her sword with her free hand. Then she held her weapon by the middle of the blade, as if she were holding a spear, and threw it right at the helmet of the giant Angel. The sword flew like missile and hit the Big Monster right in the eye. The throw was so fast and powerful that the sword went through the giant Angel's head and came out of the back — causing a big explosion at the rear of the giant's head. A lot of blood and pieces of brain were spread through air, making the giant lose consciousness and fall heavily on its side like a tall building being demolished.

Muriel's unbelievable strength impressed Glenda, who was gripped by fear and wondering what she could do to her.

"Oh gosh, she'll kill me. Of course, she'll kill me."

Muriel turned toward Glenda and saw her through bloodshot eyes. Then the warrior Angel removed some of the distorted hardware, like it was made of cardboard, letting Glenda free.

Glenda felt a lot of pain and thought the Angel wanted to kill her, but she was mistaken.

"Do not worry, brave warrior. Just come with me," Muriel said in a sweet voice.

The winged woman carefully took the injured pilot in her strong arms, and with a firm movement of her legs and wings, she took off towards the sky as the giant robot was covered in flames.

It was hard for Glenda to believe Muriel's attitude, but she was in so much pain that she didn't have any other option but to let the Angel carry her up to the sky.

Muriel flew, graciously carrying Glenda in her arms, dodging with agility the drones and other Angels.

In the sky, Glenda was mesmerized with the horrible scenery down on the ground. A scenery full of battles, explosions, pain, and terror. Also, she noticed, on the roof of Cloud Base, the last Big Monster kept trying to pierce the building with his spear, even with lots of drones shooting at him.

The winged warrior flew away from the war zone and landed in an area far from any battle. She carefully laid Glenda down on the ground behind a large rock.

"Stay here, brave warrior. You have to recover," Muriel said.

Blood and sweat mixed on Glenda's body, and her army uniform was totally dirty. Also, her face still showed her astonishment.

Muriel turned back, and when she was getting ready to take off again, Glenda caught her attention, asking:

"Hey... why did you save me?"

The warrior Angel turned and looked at Glenda for a moment.

"You helped my Messiah. You deserve all the blessings of the universe."

Then, with a strong push of her legs and wings, she took off and quickly reached the dark sky, going back to the battle.

Glenda stayed on the ground just watching Muriel fly away and refusing to keep away from the war.

"Thanks, Angel, for saving me, but the war continues," she whispered.

Using all the strength she had left, Glenda got up, and even limping, she went slowly back to the war zone.

CHAPTER 66

Inside a huge area in Cloud Base, several soldiers pointed their guns at the injured intruder who slowly took off his helmet. All of them were startled when they noticed the exhausted-looking guy was Antony Sims, one of the heirs of this place. The leader of the group, who was standing ahead of the rest of the soldiers, sounded surprised.

"Damn… Mr. Antony. Is that you?"

"Yes… thanks for the welcome, but now —"

Suddenly a weird and sharp sound buzzed from the radio on the chest of the leader of the soldiers, near his shoulder.

Antony was interrupted by the sound and noticed a camera on the leader's helmet, which made him worried about who could be watching him.

A voice from the leader's radio could suddenly be heard by everyone. The transmission was bad, but it was enough to recognize who it was — and Antony hated to know who was speaking.

"Mr. Sims, what are you doing there? It's not a good place to be right now," Yelena Orlyk's voice sounded overbearing.

"Oh, no… It's you," Antony sounded disgusted. "Ok, the radio signal is the only thing that can enter this place. It was probably my cousin's idea. So, I'm not a *radio signal*, which is why I used my motorcycle to get here."

"Be polite, Mr. Sims. These men only follow my orders. What are you doing there?"

The soldiers kept their guns pointed at Antony, who decided to raise his hands.

"Ok, ok... I can... I can end this war, but I need to get inside Cloud, and to do that, I have to get to Christine's office. That's our only chance."

"So, tell me about your plan, Mr. Sims."

"TELL YOU ABOUT MY PLAN? The world is ending out there, and you're probably hundreds of miles away from here, drinking your triple shot of whisky. I don't have time to tell you my plan and —"

The sound of an explosion echoed through the corridors, and everyone realized that something bad was coming, and coming fast.

Outside Cloud Base, on the roof, the last Big Monster finally managed to drill through the thick roof that was made out concrete and steel. The hole was small, but it was enough for some Angels to start getting inside the Cloud Base installations, even with several drones flying around and shooting missiles at the giant.

Screams that sounded like wild eagles and the sound of wings seemed to be coming closer, making the soldiers apprehensive.

"Yelena, the Angels are getting inside. I can save this place, I promise... please," Antony begged.

"He's right, Prime Minister. The noise of the monsters seems closer now," the leader of the soldiers said through the radio.

After a few seconds of silence, which seemed like an eternity, Yelena started talking again.

"Well, Mr. Sims, I'll let you go, but —"

The leader abruptly turned off the radio on his chest and said:

"No worries, Mr. Sims. We're gonna cover you. This blonde cow is so boring."

"Wow... Ok," Antony took a deep breath, "thanks, men. Keep safe."

"The east elevators are working, Mr. Sims... and you, men, be ready to defend this area."

The soldiers were getting ready to face the evil which was coming as Antony ran to an elevator. There, he propped himself on the wall and started to feel a sharp pain in his chest, in his heart.

The elevator was fast, and Antony's agony grew more and more, especially when he heard the sounds of shots and Angels' screams coming from where he had been only minutes ago.

Antony was sweating a lot, and his veins were dilated under his pale skin. He opened his coat, and the big scar on his chest seemed to be more red than usual.

"Come on... damn heart. Don't let me down now... not now."

A desperate feeling gripped his mind as he punched his own chest, even knowing that was useless.

The elevator doors opened, and Antony saw Christine's office at the end of the corridor, but his eyes could only see the way dimly.

He ran awkwardly until he got to a keypad on the wall in the middle of the corridor. He felt like he was going to faint, but he still typed quickly and closed three solid steel doors along the corridor behind him.

The sounds of Angels and gunshots seemed like they were approaching terribly.

Antony got to Christine's office. One more keypad on the wall, one more sequence of numbers which only Antony's quick mind could decipher, then he got in.

All parts of the office lit up automatically, including the wide desk and touch screen. The high-tech office seemed to be alive, more alive than Antony who struggled beyond normal to walk.

On the modern aluminum shelves, Antony looked for a Headgate, and when he got it, he instinctively set it on his head.

His heart beat increasingly faster, making him feel like fainting even more.

"Come on, damn heart... just a few more seconds, please," agony took over Antony's face.

Despite his body being almost lifeless, his mind was steadfast. In front of the wide desk, Antony typed some commands on the touch screen, making another solid steel door close the only access to the office and preparing the system to take him into Cloud.

Like a patient in critical condition, Antony lay down on the desk as it showed violent battle scenes inside and around Cloud Base.

"I hope I don't faint in the middle of the passage. Ok... here we go!"

He took a deep breath and closed his eyes.

CHAPTER 67

Kevin, a young and inexperienced soldier, ran desperately among the flames and rubble holding a rifle. Cars, tanks, giant robots and giant Angels were fighting all around him as he ran, completely lost, until he found a secure shelter behind a destroyed, overturned pickup. He was panting, and he started trembling even more when he noticed a giant Angel approaching him — a large winged warrior wearing gray and green armor and a medieval helmet that looked like the head of a locust. Kevin unlocked his rifle, took a deep breath, and pointed it at the giant. His finger held the trigger tight, but the rifle didn't shoot.

"Come on... damned gun. WORK!"

Kevin tried to unlock the gun many times but was unsuccessful.

The giant Angel, holding a long sword, approached Kevin slowly as he stayed behind his shelter with no options to escape or attack.

"Come on, gun... work, please," Kevin grew more and more desperate.

The shadow of the giant covered the young soldier who closed his eyes, feeling that his time in this war was over.

The winged warrior lifted his long weapon, ready to strike his target.

Tension and fear mixed until Kevin heard a buzz.

A missile hit the Angel in the neck, making his head fall off.

Kevin opened his eyes, astonished with the giant Angel on fire with his heavy head suspended by some fleshy fibers. The giant lost his balance and fell on top of a big pile of rubble.

The young soldier looked around, trying to see where the missile had been launched from and saw, behind him, Glenda Jackson walking slowly, limping, and wielding a small bazooka.

"I've got into the habit of saving boys. It's probably my destiny, haaaa —"

"General Glenda."

Glenda sounded sarcastic, until she felt a sharp pain on her leg. Kevin ran to help his General. Both hid behind the destroyed pickup and stayed there — crouching and protecting themselves. A breath of relief blew on them.

"Thanks, bro," Glenda breathed fast, "my leg is so sore... haaa.... to be honest, I'm sore all over."

"You're perfect, General. You saved my life. If this gun would at least work."

"Let me check."

Glenda took Kevin's gun, unlocked it, and shot up. Kevin was surprised by how easily his General got it.

"The secret is to do it quickly, with a steady hand. It's a matter of practice."

"Glenda gave the gun back to the young soldier."

"Ok, I got it," Kevin looked nervous. "You were probably the best soldier in military academy."

"Thanks, bro, but I never went to a military academy."

"Never? So, how do you know how to handle guns so well?"

"We live in America, bro. Everyone knows how to handle a gun in this country."

"Ok... I get it. So... what's the plan now?"

"We already did our best. Now all we have to do is wait... and survive."

Some Angels flew over the area, shooting arrows and spears, which made Glenda draw her pistol and stay alert, looking up. Kevin looked calmer, pointing his rifle up to the sky.

The moment continued tense, and the chances of winning this war were basically none. However, Glenda took a moment to look at the face of the young soldier.

"Hey, bro, hmmm... your eyes."

"What?"

"Your eyes...they look like my son's."

"Thanks, General... so, your son has good sight."

"Yes, he does," Glenda smiled, "keep your good eyes open. We're going to survive... I promise."

More Angels flew over the area, screaming like wild eagles as Glenda and Kevin shot their guns like there was no tomorrow.

***

On the other side of the Black Hole, Cassiel still stood on the peak of a mountain, just watching the immense army of Angels that filled the horizon.

Cassiel took his copper hunting horn and carefully looked at it in his hands, knowing well that with just one blow he could start an apocalypse and a new age on Earth. His pale face and red eyes didn't express any will to do it, but he knew that his mission was necessary.

One blow, just one blow could change part of the universe, although he knew this action didn't depend on him but on the order from someone beyond.

"My God, I hear Your voice, and I know that the time is close... very close... if it is your will, make the humans destroy the Cemetery of Sous before this last blow... please!" Cassiel whispered like a priest in prayer.

He held his hunting horn with a steady hand.

***

The American air force approached the Nebraska desert, and all pilots seemed to have the same foreboding. *Something bad is coming, and it'll be devastating.* It could be something supernatural or something from the Black Hole that spread through the air until it got every heart and mind.

Despite that, the squad continued undaunted, like a huge cavalry up in the blue sky. Thousands of well-trained pilots seemed to lie to themselves when they said something that showed hope.

A clear horizon was ahead of this powerful and worried squad, and in the middle of it, a black dot stood out right on their line of sight. The radars on their on-board computers showed a strange and terrible movement on this black area, making many pilots talk among themselves about what could be happening there. Conversations about quitting or aborting the mission echoed through the radio waves of the squad until it got to the ears of pilot Eagle 23, President Sowell.

Richard's wish to keep his secret inside the squad went against his position as a leader. He didn't have another option in his heart except to connect his headset for everyone to hear his voice.

"Dear pilots, this is the leader of the Air Force... remember what you heard in Wisconsin... we shall fight, and we shall care for each other in the battle and after the battle. Because we'll win and we'll go back to our homes... keep it in your minds and hearts. Also remember... even though we walk through the darkest valley, we will fear no evil."

Silence took over the radio waves for a moment, and most of the pilots recognized the voice of the President of the United States immediately. *Impossible... it's him... and he's crazy,* were some of the things they whispered.

"We'll see... we'll see the sun shine in the end of this story. Believe me, dear pilots."

These last words hit the pilots hard, leaving a spark of hope in their minds. They were still worried about their mission, which still seemed to be a suicide mission, but that was the right thing to do.

One of the leaders of the squad confirmed: *Ten minutes to reach our target,* and another one replied: *Ten minutes to change our lives.*

CHAPTER 68

"Mom... MOOOMM... It's me, Antony. Where are you?"

Antony finally got inside Cloud and looked for Angela inside her house, but the place was messy and it seemed to give off grief and fear. Groceries, pictures, vases, and the damn medical documents Antony had brought last time he'd been there were spread over the carpet.

Antony looked invigorated, no wounds, no burns, no tiredness, and his clothes were in good condition.

"MOM... I'm here... damn it!"

Angela wasn't there, so Antony left the home and faced another surprise.

It was night, and up in the dark sky, red and green lines blinked constantly like an immense short-circuit over the neighborhood. Several pieces of furniture like tables, chairs, sofas, and other items such as carpets, vases, curtains, and TVs were spread along the streets and sidewalks in front of the houses. Also, there was no sign of people, as if everyone had thrown out the first thing they saw in front of them and desperately ran away.

"It's more than insane," Antony said, "it's bad... very bad. My mother can only be in one place."

Antony ran along the streets toward the place where he'd always accompanied Angela for the last few years, every single weekend. He noticed all houses were abandoned with a lot of stuff spread in front of them — no happy children playing, no noise of teenagers having fun or adults taking care of their gardens, no women talking joyfully as they walked through the streets, or the smell of barbecue in the air. The environment Antony had always criticized, calling it a place so *sickeningly happy*, full of empty people, in that moment was somber, like a depressive dream. As he ran, he started to realize that his criticism was egocentric, wishing he was walking with his mother on another simple Sunday morning. Also, the worst was still to come.

"Wow... this place looks like a nightmare. What could have happened to everyone? According to the computer in Christine's office, Cloud is working well and... damn it, what is this?"

Antony saw something strange on the streets, close to the park, where the church he and his mother used to attended every Sunday was located.

Around the church Antony had built for Angela, thousands of people — adults, kids, teenagers — knelt, with their faces down, crying and trying to pray. They were spread out on streets, sidewalks, and especially in the park, as if they were a huge sea of people, and the church was a little island in the middle of them.

"Oh... they are probably feeling so sad. But I don't have any other option."

Carefully, Antony went toward the church, walking among the people who didn't notice his presence. With each step, he heard whispers of sadness, shortness of breath, attempts to pray the Hail Mary or the Lord's Prayer — which sounded odd — cries, and groans.

They seemed to have forgotten the concept of death, the idea of dying someday, and in that moment, by instinct, they went there, to try to apologize for something they didn't know how to explain but which made them feel bad. Maybe they felt shame for exchanging their beliefs and religions for an illusion, a life full of pleasures, or they felt frightened, finally realizing that everything could end. The doubt and fear of not knowing where they would go made them act this way.

Antony felt like he was floating on an ocean of doomed people as red and green lightning ripped the dark sky. *Maybe the vibrations from the Black Hole or some kind of energy emanating from the Angels had affected them.* This was one of the ideas that drifted through Antony's mind. Another idea was the certainty that they were just consciousness, a virtual reflection of their emotions and will. If they had been in a tangible world, they would be acting the same way.

I.G.W.T

It was impossible to see the grass on the ground of the park; only the oaks and the church rising among the kneeling crowd could be seen.

As Antony approached carefully, he watched the sculpture of the archangel Michael drawing his sword. Coincidence or not, he had an epiphany when he realized this angel looked like those which he had faced the last few days, and another surprise hit his heart and mind: the initials I.G.W.T above the sculpture, the ones Angela had asked him to write on the pediment in front of the church.

"In God we trust... yes, Mom, it's so obvious. Especially coming from you," Antony got emotional.

The doors of the church were closed, and the crowd stopped on the steps. *Why didn't they go inside? Perhaps they are ashamed to enter.* Antony tried to understand the reason why people hadn't entered as he opened the door carefully.

Inside, it was almost empty — a gloomy place illuminated by the glimmer of the night outside. Everything looked the same: walls adorned with small sculptures of angels, cherubs, and golden details in a baroque style. The large altar in the back was hidden by shadows, as the candlesticks with red candles were extinguished. *It's dark and silent like the end of the world,* Antony thought.

He was overwhelmed by fear and apprehension as he walked carefully among the pews, his eyes focused on the chapel in the corner, where the replica of the Virgin and Child was set. He'd never imagined he would miss seeing this sculpture. Mary holding the baby Jesus as they looked at each other brought a spark of hope through flashbacks of sweet moments, when he'd come to the chapel with his mother. In front of this beautiful sculpture, there was another, not so beautiful one: the gray stone book on a four-feet stone stand, and on it, a high-relief infinity symbol.

"You have to be here, please. You have to be here," Antony whispered.

A strong feeling of relief hit him when his eyes noticed a small old woman passed out in front of the gray sculpture of the book on a stand. Her silver hair was covering her face, but Antony had no doubt who she was.

"Mom... MOOOMM!"

Antony ran toward Angela as if he had to apologize to her before the end of the world. He was filled with anxiety as he crouched down and carefully took his unconscious mother in his arms. His eyes were wet, and his lips trembling.

"Mom... It's me, Antony... your Tony. Please, wake up."

Angela's eyes slowly opened, and she almost couldn't believe who was with her.

"Tony! It's you... my son... my only son."

"Mom!"

Angela wished her son had forgotten her mistakes and hugged him tight. Her wish was granted, and she was sure that everything was well between them because she felt love in this long hug.

Tears rolled down both faces, and the trembling voice of Angela murmured.

"Sorry... I'm so sorry. I was a horrible... horrible mother. Sorry, my son —"

"No, you weren't. You are an amazing person. You've always been amazing. Your father... my father... Grandad Allan, he told me. I saw him... and you... holding me in your arms when I was a baby... saying you loved me. I saw it."

Both stepped back a little to look at each other.

"You saw it? Me and my father? How?" Angela was very impressed.

"We don't have much time, Mom," Antony smiled, relieved. "There is something very important I have to do right now."

Mother and son got up and stayed there, calmer from that moment, in front of the sculpture of the book and close to the Virgin and Child.

"I came here to pray and... I don't now... suddenly I fainted and... anyway... things aren't going well outside, I assume."

"No, they aren't."

"So... you came here to fix this situation. Is that it? You can fix anything, I know that, Tony."

Antony had another epiphany, which made him look up and smile with his eyes still wet. Angela thought this sudden behavior was odd coming from him.

"What is happening, Tony?"

"I just remembered... today I met a wise man, and he told me something that makes perfect sense now. And it's so incredible," Antony looked impressed. He told me: 'We're just where we have to be, doesn't matter if our actions were good or bad. These actions always lead us to do the right things in the right place.'"

"Hmmm... I still don't understand you, my son?"

"But you will. Look at that."

Antony took a deep breath, preparing himself to do something he had wanted to do for a long time. Then he reached his hand to the high-relief infinity symbol on the stone book, spinning it slowly with his fingertips — like the symbol was a safe lock. After two spins, something that seemed like magic happened. The book automatically opened slowly, as if it were a machine made of stone. It opened completely — revealing something mysterious. In the middle of a gray smooth surface, a low-relief open hand shone in an intense blue light that illuminated the faces of Antony and his mother.

Angela was mesmerized by the secret revealed by her son and the intensity of the light. It was possible to see thin, blue veins shining under the hand.

"It's... it's so beautiful," Angela was intrigued. "But what is it?"

"It's something that I hid here some years ago when I built this place for you. If someone presses this hand — not any hand but the hand of a special person — it'll launch a kind of algorithm that will destroy the programs that keep the consciousness of everyone who lives inside Cloud. In short, it's a kind of button that can turn off all of Cloud... I called it... Key."

"Key! Oh... so, you'll turn this place off... including your mother. Right?" Angela was apprehensive.

"Not exactly. Look at Key."

Antony reached his hand close to the blue hand.

"Its shining hand is small. My hand is too big, but yours... yours... is perfect."

An immense wave of different emotions crossed Angela's heart — a mix of responsibility, fear, and hope. Something impossible to explain in this short and decisive moment. She was frozen, trying to find words to describe it, as Antony confirmed her doubts.

"It's you, Mom... the only person who can save everybody here and stop something terrible from happening to our world."

"All this time you hid it here... how can you hide something so powerful like this and... why me? It's such a great responsibility."

"Did you think I would create something to destroy this place and give myself access to it? No, I wouldn't. I know myself, I know my temperament, my imperfections... every single day I distrusted myself, my actions... but you... you're the best person I have ever met. The most honest, able to love unconditionally. I truly wish we had more time to talk about this, but no... the decision has to be made now, and it's yours... the only one I trust."

Angela stayed in silence, just watching carefully the open hand shinning its blue light that reached the roof of the church and covered the Virgin and Child in front of it with a soft blue light. Angela's wet eyes focused on Mary and Jesus for a moment as the sounds of the people crying outside still echoed in the distance.

"But... If I decide not to press this hand?"

"It's your choice... to me, your choice is the best choice."

"And if I end everything... Will I see you again?"

"I... I don't know."

Tears fell from Antony's eyes as a deep sadness gripped his heart. Angela felt this pain corroding him and caressed his face.

"No worries, my son... Do you hear the people crying outside? They cry because this world... this world, made by my father, never gave them this one valuable thing, something that made us humans."

"What would that be?"

"That is... faith... I'm sure we'll see each other again, and you must be sure too."

"Mom... always saying the right words even with the world almost ending."

Tears intermingled with smiles in both faces, expressing pride. A tight hug joined mother and son in a moment they wished could last forever.

"My mom."

"My son."

Antony closed his eyes and held his mother in his arms, keeping his chin on her head. Angela kept her head comfortably on her son's chest.

The blue light from Key covered mother and son, creating a beautiful blue light sculpture in the darkness of the church.

If this moment were to last forever somewhere in the universe, Angela didn't know, but she kept Antony's arms around her as she carefully placed her hand over the low-relief shining hand.

Her hand fit perfectly.

CHAPTER 69

"General Glenda... I'm out of ammo," Kevin said, breathless.

"Take this pistol. I took some on the way here."

Glenda and Kevin were still hiding behind the destroyed pickup.

Angels and giants kept fighting as if it were the beginning of the battle while humans and their giant robots suffered because they had almost no strength, bullets, or hope left.

Glenda loaded the last magazine into the pistol and got ready for what she felt would be the last offensive. However, when she saw the gloomy area in front of her, fear took over her mind. Three giant Angels approached slowly, making the ground shake as hundreds of others flew over them.

She swallowed a little of her own blood and unlocked her gun. A prayer echoed in her mind when she heard Kevin's scared voice.

"What now, General Glenda?"

"Just keep fighting... and keep your faith."

***

On the other side of the Black Hole, Cassiel looked up at the immense army of Angels ready to start the upcoming apocalypse. He kept standing on a mountain peak with a long copper hunting horn in his hands.

His words sounded severe.

"The time has finally come... I can hear my God whisper in my ears."

Cassiel slowly lifted the horn close to his pale lips, ready to blow and command the final attack against Earth.

***

Up in the sky, President Sowell led the immense fighter aircraft squad over the Nebraska desert, very close to the dark area in the middle of it. At this point, every pilot knew that the President of the country was leading the mission.

His next orders sounded brave:

"Dear pilots, keep your guns unlocked and stay ready to attack, have no mercy on the enemies of our land."

All fighters obeyed immediately, the tension overwhelming every heart as the dark sky became more and more obscure ahead of them.

"Dear pilots, in less than one minute, we'll get in the war zone... one, two, three... OH LORD, WHAT IS THIS... ABORT MISSION, ABORT MISSION... EVASIVE MANEUVER, NOW!"

Suddenly, the impossible happened right in front of their eyes. An explosion, a light, a sun, a flash... The roof of Cloud Base lit up, and an immense beam of light arose from it, hitting the dark sky and illuminating it completely. Up there, a vast swirl of light that looked like a cyclone appeared. So vast that it easily made the Black Hole disappear.

Glenda and Kevin stayed behind the destroyed pickup, understanding nothing about the intense light that flooded everywhere they looked.

Cassiel was startled when he heard a sound that he understood as the *roar of his God*, a harsh sound that made the gray sky lighten up. He immediately put his hunting horn away from his lips and noticed the Black Hole was vanishing behind him.

"My God finally decided... It is done."

Then, his worried expression changed to relief.

On the war zone, all kinds of Angels stopped immediately and just looked up at the immense cyclone spinning up in the sky, and all the soldiers were frozen, mesmerized by the spectacle. A cold wind blew, spreading sand and putting out the flames.

No more battles, no more shots or explosions, just light and contemplation.

Glenda and Kevin were so scared that they instinctively hugged each other like mother and son. The light was so intense that it was hard to open their eyes and see what was going on around them.

"What is happening, General Glenda?"

"I don't know... just stay with me."

Glenda risked opening her eyes, but it took a while for her mind to believe and assimilate everything she saw. All of the giant Angels were floating, rising slowly to the sky. The corpses and blood spread on the ground were also rising to the sky. All of the corpses and traces of blood were transformed into a golden light as they went up to the sky, like magic, like the immense cyclone above wanted to remove any traces of the death and pain from the war zone.

Up in the sky, the drones flew away from the beam of light, along with the immense squad that was flying around it, while the Angels flew into it, spinning like a flock formed by thousands of birds. But the most incredible was still to come.

Over Cloud Base, millions and millions of golden, bright spots arose like stars, leaving the building and rising to the sky, toward the cyclone that kept spinning, surrounded by white clouds.

Every aircraft fighter kept flying around the illuminated airspace as their pilots just watched, impressed, the bright spots and their trajectory until they went into the new gate. For each pilot, the immense light beam filled a large portion of the sky with a piece of the galaxy.

One of the pilots decided to call President Sowell.

"Mr. President, what will our next maneuver be?"

"I… I don't know. Just keep watching this amazing scene. You will probably tell your children about this… and your grandchildren too."

Between Cloud Base and the new hole, that looked like a cyclone, a moving constellation was formed, and the Angels flew among the stars — like they were swimming in an ocean made of light that flowed, free and continuously.

Glenda was so mesmerized that she instinctively stared toward the constellation that was leaving Cloud Base up in the sky and came to a conclusion: "Souls… each bright spot is a free soul, free to go to heaven". In her mind, stars and angels were dancing to a well-composed symphony.

"Maybe this is how the universe was born," Glenda whispered.

Bright spots, stars, souls, spirits… it didn't matter what people up in the sky or on the ground imagined. The war ended, and a fresh wind of relief enveloped everyone. A good feeling, that seemed to come from nowhere, arose in all hearts and minds.

All bright spots and Angels finally went into the cyclone after a few minutes that seemed endless. This new gate closed, leaving a desert with a bright blue sky and many curved clouds.

Glenda and Kevin got up and left their shelter, watching mesmerized the environment around them. No more battles, no more shots, explosions, or screams. Just a desert of rubble, destroyed war machines, and sand. The Cloud Base building was stable in the horizon. A sad scenery at first sight, but in peace.

Up in the sky, the immense aircraft fighters' squad still flew over the desert, trying to find any trace of the Angels, though quiet reigned over the desert and they couldn't find anything.

Some soldiers appeared amid the rubble, still amazed at what had just happened. Some of them said that the best option was to go to Cloud Base and look for some medical treatment. Everyone agreed, then they slowly went toward the building.

Glenda was badly injured and tired, but in peace, leaning on Kevin's shoulder as she walked. She finally breathed, pleased, with no worries about the next attack. As she watched the white building at the end of the horizon, some words instinctively left her lips:

"He did it... the little chicken did it. It's unbelievable."

"Little chicken? What do you mean, General Glenda?" Kevin sounded confused but was smiling.

"Sorry, bro, little chicken is how I called... humm, forget it. It's a long story."

"Well, we have a long way to walk, and the General can't walk without my help, I guess."

"Ahahaha... Ok, bro. You convinced me. Let me tell you from the beginning. I was living my simple life in my community when a guy appeared from the sky..."

***

In another world, far away from Earth, a vivid landscape formed by great and beautiful green mountains appeared since a Black Hole had vanished from the gray sky. At the end of the horizon, a sun rose, coloring the sky with orange and blue tones.

Thousands and thousands of Angels flew over the mountains, contemplating the new-colored land, dancing up in the sky, celebrating a new age full of good expectations. One of them couldn't hold back her emotion seeing her home beautiful again. Muriel. Tears rolling down her face and a contagious smile seeming to shine on her lips.

"My Messiah did it, yes, he did it... God was right in believing in me and him."

The sadness and the dingy vibes disappeared completely, and everywhere the colors seemed more and more lively.

Muriel landed on the top of a mountain where she had, several times, heard advice and teachings from her master, Cassiel, who looked a little upset.

"Cassiel, my master. It is time to celebrate."

"I know... my dear warrior... You were amazing in your mission. But me... for a moment, I almost destroyed the plans of our God... for a moment, I almost blew the hunting horn and let our army invade the world of the humans... Our God stopped me in time," Cassiel was always demanding about his own altitudes.

Muriel approached her master and caressed his disappointed face. Her eyes were adorable and serious at the same time, as her face only showed the affection of an obedient daughter.

"Do not worry, my master... maybe... maybe our God wanted you to feel doubt and a little guilt in order to eliminate any trace of ego that still lived inside you," Muriel winked.

"Yes... yes. You are right," Cassiel smiled, "everybody has something to learn, no matter if you are a beginner... or a master."

"God knows how and where everything must happen... you taught me this lesson, my master."

"I remember... Also, for a moment I doubted you and your Messiah... your brave Messiah. He was the right choice and you... the right warrior."

For a moment, the master became a learner, and a wind of peace invaded his consciousness. Cassiel touched Muriel's hand as a gesture of gratitude when he noticed someone landing behind her.

"As you said... it is time to celebrate... especially you, my lovely warrior."

Muriel noticed that her master was looking behind her, so she turned around to gaze at someone very special.

"Abaddon... my love."

"Muriel... my *favorite opponent*." Abaddon was glad.

"My dear warrior, go and enjoy your glory," Cassiel said.

Tears intermingled with smiles on her face as Muriel hugged her master tight. Then she went away, directly into the arms of her love.

"My love... my dear Abaddon, our mission is over."

"Our mission? No, your mission. You dedicated a life to it. Now you deserve a life to enjoy its fruits... before another one appears."

"Another mission?" Muriel looked doubtful.

"Remember, we are warriors. The universe always needs us... but not right now," Abaddon smiled more and more.

"Yes... not right now."

Muriel's face seemed to shine with happiness as the two brought their lips together.

Two Angels kissing in front of a sunrise that shone behind a vast green mountain chain as thousands of Angels flew in a bright blue sky.

This side of the universe had been saved.

CHAPTER 70

In Cloud Base, Yelena Orlyk and Christine Sims looked powerful as they walked through the wide corridor accompanied by a group of several armed soldiers. They were going to the office where, days before, Christine was able to control and watch everything inside this building. Yelena, as always, looked powerful in her black pantsuit that accentuated her height and her athletic body, unlike Christine who looked like a child cosplaying as the Prime Minister, wearing the same kind of clothes. Both women expressed a mix of seriousness and fury on their faces, as if all their plans had failed — and they really had. This group approached the office's locked door, and by its side, a security guard tried to unlock it, fixing the keypad with some tools.

The group stood in front of the office's door, just waiting and making the security guard tremble with nervousness. However, after a few minutes, he managed to fix the keypad, and the wide door opened completely. Then, everyone got in immediately.

Yelena and Christine went ahead, and the soldiers followed right behind, wielding their guns. Everything looked normal inside, except for the person behind the large desk and the smell of alcohol in the air.

Antony slept in the large armchair, and on the desk, some bottles of whiskey and soda, most of them empty, were spread. His sleep was so heavy that he didn't notice the presence of the people inside until Christine started screaming furiously:

"ANNNNTONY... WHAT THE FUCK DID YOU DO?"

Everyone was startled by her scream as Antony woke up, calmly and completely drunk, opening his eyes slowly until he saw Christine, who looked like a furious hunting dog. Yelena kept her expression very serious, and the soldiers were pointing their guns at him.

Antony sounded sarcastic and smiled goofily the whole time.

"Heyyyy, guys hehehe... you're late to the party," he grabbed an empty bottle of whiskey, "well... the drinks are over, but there is some water in the fridge, I guess... Wow, my head feels like it's going to explode. That's why I always preferred beers."

For a moment, Christine looked like she was going to attack her cousin.

"I'll explode your head if you don't explain what you just did. None of the computers in our corporation are working, none of our satellites are working, and worse yet... Cloud is empty... EMPTY... you transformed the enterprise of our family into an immense hard disk full of nothing."

"Wow... maybe I stumbled on a power cable, I guess. It happens in the best families, hehehe... oh, my head."

"Antony, this corporation... this place is our life. You just destroyed everything — our technology, our wealth... including Cloud Base Two. I don't know how and... What are we going to do from now on?"

"Well... you better you update your resume. Do you have any experience washing dishes?"

"HAAAAAAA... I CAN'T TAKE IT."

Christine Sims left the office more furious than when she had arrived as the soldiers stayed frozen, pointing their guns at Antony and thinking this situation was very odd.

Yelena took the floor and looked directly at Antony, trying to pretend she was in control of everything.

"Mr. Sims, do you realize the tragedy the world will go through because of your attitude?"

"Yes, I know... I know very well. I mixed whiskey with soda, and it was a big tragedy for the world. The taste was horrible."

"Mr. Sims," Yelena looked like she could explode at any time, "I remember you promised me you'd save this place."

"And you believed me, hahaha... well, the last one to leave the office, turn off the lights, please."

Antony closed his eyes again, ready to go back to sleep.

The Prime Minister had flames in her eyes when she turned to the soldiers behind her. Her powerful voice echoed through the office and the corridor as the soldiers listened to her carefully.

"Men, this is the most important order that you'll ever get in your lives. Arrest Mr. Antony Sims immediately. He is humanity's public enemy number one. I, in the name of governments from all nations, am accusing him of crimes against humanity, including inciting a terrible war that damaged millions of —"

Suddenly, a strange noise caught Yelena's attention, interrupting her speech. Then, she turned around again and saw Antony snoring and drooling as he slept, hugging a bottle of whiskey, in the office armchair.

All Yelena wanted was to grab Antony by his neck and strangle him until he agonized to death, but her class and position would never allow her to act this way, even knowing her career would collapse. She took a deep breath and quickly left the office, shouting her last order:

"I give up... I give up, do whatever you want with this damned creature."

Yelena vanished at the end of the corridor as the soldiers lowered their guns and looked at each other, not knowing what to do.

Antony slept heavily, as if he had done nothing, but his consciousness knew very well that he had changed the world.

CHAPTER 71

The following days, after what the media called the *Sims Reset,* in reference to all the events caused by Cloud's destruction, were marked by deep transformations in society, especially in its mentality. Actually, the term *Sims Reset* was a euphemism to disguise the bankruptcy of the most powerful corporation in the world, which had become only a few shares no shareholders wanted to trade.

Christine Sims, other members of the family, and loyal investors desperately tried to restart Cloud's system but were unsuccessful. Algorithms, codes, commands, millions of chips installed in the brains of everybody around the world — everything connected with the Sims Corp became useless. The algorithm called Key, created by Antony, contaminated every computer like a virus and deleted all their memories, turning all of this great corporation's technology into an endless amount of garbage.

The only person who could fix this new and decadent reality for Sims Corp was its heir, Antony. However, he hadn't been paying attention to the desperate orders from his family members, politicians, investors, shareholders, and business partners for him to restart Cloud. All he said was, "I have no idea what happened," always smiling sarcastically, even with everyone promising him riches and power like in the past.

Antony became an *iceberg man*, like his grandfather, but acting in the opposite way, always smiling and not giving a damn about his family's riches.

In this period of the *Sims Reset*, Antony was taken to a special jail in the Capitol Building, Washington, and there he was confined until authorities decided the effects and directions the world would take, having the old concept of death again.

The media went crazy, creating more and more news and fake news about the end of Cloud. The social network was out of control, releasing millions of speculations and opinions around the world, making millions of people live in a place where everyone believed in nothing, even in its leaders.

Economies in decline, unemployment in rich and poor countries, violence, and the loss of hope. Citizens around the world felt like orphans of their leaders and media. A kind of *anarchist wave* was ready to begin, and the fearful governments had the idea to blame someone, and there was only one person to blame: Antony Sims.

A great campaign, led by the media controlled by politicians and powerful corporations, started showing Antony as the world's main enemy, guilty for all the problems around the planet. The guy who had been shown as the world's sweetheart by gossip shows and news after the tragedies in Nebraska was now shown as evil, the one to blame for all the unemployed, deaths of starvation, and acts of violence. Also, the governments were shown as the only salvation, and their leaders started to see this situation as a new opportunity to retake control in declining societies. But these leaders had forgotten something: while many people were being manipulated by the media, most of them still distrusted showing their opinions, theories, and reasonings on social medias. Also, the book written by Antony, *Cemetery of Souls*, was widely shared because everybody wanted to know why the heir of the richest corporation in the world had decided to destroy all his wealth. The answer wasn't in this book, but one piece of information in there drew a lot of attention. Something the media hadn't shown or given any opinion about: the fact that, in ten years, nobody had been born on the whole planet, and according to *Cemetery of Souls*, the reason was the existence of Cloud, as if this virtual world had caged the souls that should go to heaven. A ludic explanation but intriguing, when the reality was analyzed.

The apparition of winged creatures in Germany, France, and the United States was reported as a lie, conspiracy theory, fake news, or manipulated images. The military was instructed not to speak about the incidents that involved Angels, with the risk of being arrested in case they leaked any information. However, the people were exhausted from being cheated by those who controlled the media. While the main communication vehicles treated the topics about Angels as urban legends, hallucinations, and even hacker groups promoting disorder, throughout all countries, a new consciousness arose, and the majority of people realized that a great phenomenon had happened. Millions of people believed the scenes of winged beings that were broadly shared on the internet, and the destruction of Cloud just confirmed that something great had changed the world.

***

The *Sims Reset* was almost completing one month, and Antony was still in prison. Alone, in the hands of the worst authorities in the world, but calm and always repeating in his thoughts his favorite quote: *We are where we have to be, doesn't matter if our actions were good or bad. These actions always lead us to do the right things in the right place.* In his cold and lonely jail cell, he remembered his two mothers, Nivedita and Angela, his new friends Glenda, the guys from community, President Sowell, old George, his own grandfather, and a mysterious woman called Muriel — a being who had appeared in his mind since he was a child and who had finally vanished from his dreams. His days in jail were marked by long hours in deep sleep, with no more nightmares about an Angel making his head explode.

However, outside the jail, protests against Antony were still ongoing, orchestrated by people who had been manipulated and were now demanding Antony's death — a sentiment that was enough for powerful individuals to take action.

Attempts of murder were secretly planned by politicians and powerful businessmen, but they were abandoned for the slightest hope that Antony could retrieve his actions. Also, the WPU (World's Peoples' Union), a global organization of government leaders and their bureaucrats who had decided the course of humanity, threatened him with the death penalty, accusing him of killing billions of people in Cloud — because according to this organization, these people who had been living in there were considered human beings. About this indictment, the heir of the bankrupt Sims Corp just answered, smiling: "Well, here we go, let's see what really is on the other side of life."

Powerful businessmen and other politicians didn't have any other option except to support the idea of WPU and let it apply this terrible penalty against Antony according to the universal laws that had been created, of course, by WPU itself. However, this kind of authority still belittled the population's opinion.

Everything was ready for Antony's death; all news and TV shows confirmed excitedly the horrible sentence for someone who, according to them, was the reason for all the evil that was plaguing the world. Marches celebrated the decision; many well-paid artists wouldn't stop bad-mouthing the guy who had tried to destroy the world but didn't manage to. Political leaders used all their power of marketing to promote themselves as the new salvation to decadent societies, societies which they had ruined for a long time. *A fresh start requires the death of evil*, that was the slogan, and this *evil* had a name, Antony Sims.

He was taken to another jail in one of the major office sites of WPU in New York for every government authority to confirm their verdict, including the new President of this intergovernmental organization, Yelena Orlyk, acclaimed as the responsible for capturing the world's public enemy number one.

The former Prime Minister of Europe had craved this position for a long time, and the choice of her name for this office was unanimous and was also celebrated by the media. Finally, she felt more powerful than the President of the United States of America, and her influence among civies, militaries, and politicians became great around the world. Of course, she used this preeminence and greatness to attack Antony and confirm authoritarian politics, especially on the internet, especially on social medias, where the people still had a little freedom to talk and share different ideas. Yelena had the nations in her hands and power to manipulate all countries her own way. *The events in Nebraska were not so bad*, she thought, as her leadership was respected more and more.

Meanwhile, Antony was kept in jail, calm, quiet, and always smiling. "I am where I have to be," he repeated to himself, just waiting for those who thought they ruled the world to kill him. Until something amazing happened… a true and great reset.

***

A miracle, a new age, a boom… actually, a baby boom.

On the thirtieth day after the *Sims Reset,* a phenomenon affected the whole world. Thousands of women in different countries were pregnant, and in the following days, the number of pregnant women just increased, causing a great euphoria. The media went crazy, the people on social medias were wild, and political leaders who had used the chaos to promote themselves became desperate.

The time came for a terrible truth to be admitted by the authorities. The fact was people hadn't been born for ten years, and it'd been kept in secret by the governments with the help of the media. Sterility had affected every social class and ethnicity of society: rich, poor, middle class, blacks, whites, Asians, indigenous peoples, those who lived in modern urban areas or in the calm countryside. Everyone had been affected, and even powerful people from the media, politics, and big corporations were emotionally impacted when they realized their wives and daughters were pregnant. "We lied, we hid a horrible problem from humanity, but we can say sorry and help to build a great world again," the president of a big channel declared. "Humanity was dying, and I was part of this mistake because I was just pursuing profit," announced the CEO of a telecommunication corporation, and other authorities, presidents, and CEOs of companies admitted to keeping this secret while humanity had stopped being born. Also, something contributed to these declarations: a book written by Antony and its ideas.

The Golden Box... one of the tales of *Cemetery of Souls*, written by Antony a few months prior, was understood as a prediction of Cloud's destruction. In this tale, the souls of people living in a medieval village were put inside a Golden box, which was a representation of life in Cloud. Once these people heard an Angel trying to break the box from the outside and then stopping without any reason. They were afraid to leave, but a messiah inside broke this kind of jail, allowing the souls to go free. Then they became human lives again and remained adrift, swimming in a huge blood lake for days until they got to solid ground. After rebuilding their village, they were ready to build another golden box, but something unexpected happened. A woman finally got pregnant, and the fact, forgotten for a long time inside the Golden Box, was celebrated and these people never wanted to build another Golden Box again. A simple and weird tale, but when it was linked with reality by people around the world, *Cemetery of Souls*, with its fantasy tales, became a book of prophecies with many stories that showed how Cloud, for many years, had been harmful to humanity.

Protests and more protests filled the streets of the main cities of all continents, just as the population flooded the internet with only two subjects: Antony's freedom and the wrath about the long silence of the media and governments regarding the human infertility. Channels and media groups tried to calm the population down with more lies and fake news — which had a contrary and terrible effect on them. Furious crowds protested, surrounding their headquarters, studios, and offices. An uncontrollable wave of people vandalized and set fire to these companies that had profited for decades deluding the population. The United States of America, Italy, England, Egypt, Angola, Brazil, Chile, Australia, China, Russia, New Zealand, Mexico, Malesia, Saudi Arabia, India, France... the list of countries where the outraged population took to the streets seemed endless. The army and police forces joined the protests too, because they also had families and were impressed when women finally got pregnant.

The world lived in a kind of anarchy. There weren't any more media, governments, corporations, or military forces leading people or influencing them. There was just a will, a great will to restart life in society, a new will that no one knew how to explain yet. A confusion, a doubt, a pair of eyes focused on a deserted horizon, a kind of worldwide feeling that whispered, *Go with hope*.

Nine months after the *Sims Reset,* the first baby after ten years was born at the end of January, more specifically on January 25th. Her name was Wěi, a cute baby who was born in Wuhan, a city located in China. On this same day, about one hundred thousand babies were born, and in the following days, more and more. This period was nicknamed the *Aquarius*

Age, according to the zodiac calendar, and caused a sequence of global events like a colossal *domino effect*.

With more babies around the world, more fathers pursued the best jobs to care for their kids. Governments and corporations heard the cries from the population and invested more and more in kindergartens, schools, and especially in security, because in a few months, the streets and parks would be full of children having fun and attentive mothers. For this same reason, great programs for cleaning and renovation of neighborhoods were executed. Teachers, cops, nurses, builders, doctors... more jobs were appearing with the goal to care for the new families in all countries.

Also, this new era brought people a wish to know how to lead their lives after such a long time without facing the concept of death. The right answer was impossible to reach, but there was something that could help: religion. Christians, Buddhists, Islamists, Jews, Spiritualists — all kinds of religious institutions — restarted their influence on people's mentality and education. Churches, temples, mosques were attended again, welcoming an orphan population with lessons and teachings about how to care for each other, especially their kids.

Polls, surveys and speculations about the return of Cloud made the people even more furious. Protests and turmoil in front of government head offices and politicians' homes cried violently for the complete destruction of this system. Many powerful people were bullied and assaulted for being suspected of organizing a kind of *New Cloud* and recreating the business of virtual heaven. However, nobody wanted the return of Sims Corp and its invention. Families pressured hospitals to not install chips on newborn brains, and this order was immediately accepted. This new generation would grow up free from any attempt of control. The last months had proved that humanity didn't worry about death if they could celebrate life.

Eleven months had passed, and Antony was still in jail in the Capitol Building, without any detailed information about what was happening outside. Politicians, former investors, and shareholders of Sims Corp finally gave up on trying to convince him to restart the Cloud system. *How can I blackmail a guy who was the richest person in the world and exchanged it for a cold jail?* they thought. Like a monk inside a cave, Antony lived every single day in silence, being fed tasteless meals in this prison, doing some physical exercises like push-ups and sit-ups in the small space available, belittling with a smile anyone who asked for a meeting with the intention of talking about rebuilding Cloud, knowing that he could be murdered in there anytime.

Sometimes, from his cold and silent jail, Antony had the impression of hearing a crowd murmuring outside, a faraway buzz that on some days sounded louder. He thought his mind had started to play tricks on him, but the fact was a huge crowd had been camping around the Capitol Building for weeks — especially after the babies started being born. The people cried out for the authorities to free him; however, the media still tried to portray him as an enemy and a danger to society, while the people saw him as the savior of humanity, making the quote *Free the one who saved our kids* the most shared topic on the internet.

A wave of people arrived in Washington from many parts of the world. More and more people gathered around the Capitol Building, and it didn't seem like they would leave soon. The crowd became so out of control that the leaders of WPU secretly took Antony to the headquarter in Manhattan, then to the one of its major offices in Vienna, Austria. There, Antony was put in another jail, similar to the one in the Capitol Building, and once again another crowd of people camped around it, waiting for the authorities to free him.

Despite the fact that the population around the world was going through the *winds of change*, full of good perspectives, there were still powerful politicians and bureaucrats, most of them members of WPU, as well as businessmen who never let their power escape their hands. Also, a strong feeling of revenge pushed them to take actions against Antony, as they felt the threat of losing their influence over society. *How did a spoiled rich guy dare to seize our power?* they cried out in their secret meetings, trying to change the course of the world again. The arrogance of this little group who insisted on seizing control of the world seemed inexhaustible.

Day after day, more people arrived and camped around the major office of WPU in Vienna, just waiting for the day of the trial, the day when the members of this global organization would decide Antony's fate. Tents and motorhomes on the streets and parks transformed Vienna into a kind of huge hippie camp, organized among boulevards and buildings of different styles like baroque and neo-gothic. Hotels and backpackers were full of Antony's supporters, and, on the improvised camps, some new mothers showed their newborns like gifts to inaugurate a new age.

The day of the trial was approaching, and Antony just waited in his new, silent jail for this decisive day. His future would be defined by vengeful powerful men meeting in mirrored buildings and the clamor of a brave crowd on the streets and the internet. Freedom was almost impossible, and the death penalty seemed the most probable given who controlled this situation.

***

One year went by, and Antony finally faced the wide and long corridor that would take him to the WPU assembly where its members, bureaucrats, and diplomats from many countries would give their verdict. To these people, the opinion of the population outside and all over the internet was worth nothing.

Antony wore a gray prison uniform and looked more fragile and thinner, especially because of the handcuffs around his wrists and the four large guards who followed him all the way. His face looked serene, he had a thin beard that matched his long oily hair tied up in a ponytail — an appearance that made him resemble an inmate and a monk at the same time.

The defendant and the guards crossed a wide set of double doors, opened by two other guards, and entered a huge chamber that looked like an ancient Roman amphitheater. This luxurious place had a high ceiling, and its walls varied between white, blue, and green colors. Several massive high columns supported the white rounded ceiling adorned with golden lines, which formed a wide world map. In the back, an imposing, empty podium seemed to be waiting for the newly sworn-in leader of WPU. Flags from all countries of the world hung around the chamber where about three thousand members of WPU waited for the defendant. As Antony walked, he observed the luxurious building and all the cold people around him. Then, some thoughts arose in his mind: *These people feel like they're gods. They built a place that looks like an Olympus and wear expensive clothes to feel like they're able to decide how we have to act and think. They aren't gods. True Gods are the highest expressions of the best of what we have within ourselves; they are the references of good things, they're what we have to follow even knowing we'll never be like them. These people aren't gods, and these flags only show their fake worries about the world. They pretend to respect the differences between countries, but what they really want is to have just one flag to represent the whole world, one population thinking the same way... the way they want, the way to keep their position in this fake Olympus... forever. They can criticize and judge everyone at any anytime. But who can judge them? The answer is nobody... because they always camouflaged themselves in a kind of varnish of kindness that hides their bad intentions. My destiny can't be in their hands, I am sure about that... hmmm, good thoughts for someone who is about to get a death sentence.*

Antony and the guards walked up to a simple podium in the center of the chamber as the public looked very serious. Each pair of eyes seemed to be bloodshot and furious with the fragile guy who looked very calm for someone who was going to die in a few days or hours. The guards walked in front of the podium, and after inspecting Antony, they left him standing in front of the main podium located a floor above.

The public in the chamber whispered fiercely among themselves, trying to imagine what would be the final verdict voted throughout the year by the members. The death penalty was their preferred choice, but it also depended on the opinion of the population outside and the conclusion of the new main leader of WPU.

Rumors had reached Antony's ears that the former prime minister of Germany and president of the European Union council, Yelena Orlyk, was the new leader of WPU, which was obvious to Antony because who else would better use a tragedy to get more power?

Antony waited patiently for Yelena to appear and declare his death penalty with her typical sarcastic smile that always seemed to have a little blood dripping from the corner of her mouth.

We are where we have to be, doesn't matter if our actions were good or bad. These actions always lead us to do the right things in the right place, George's words still echoed through Antony's mind. It didn't matter how furiously these people looked at him; it didn't matter how many hours Yelena would give him to live before giving him the death sentence. *At least I'll have a real death, not a death that takes me to a fake heaven*, Antony thought.

Time seemed to pass slowly, and the buzz of whispers from the crowd never stopped, but suddenly Antony noticed that the buzz didn't come just from people inside but from the outside too.

The sound of heavy double doors being opened came from the back of the imposing podium, and Antony, serene as ever, expected to see a tall, blonde woman full of herself, declaring in beautiful words that he had tried to destroy humanity and confirming his death sentence. But Antony's calm changed into surprise... a good surprise.

In that moment, Antony noticed that the imposing podium was a little low to adapt for a man who was approaching in a wheelchair. Those green eyes behind round glasses, the typical gray suit and red tie — the look of President Sowell was unmistakable.

Actually, the last months were very important for WPU's members survival. Antony was in the middle of an arena, an arena full of vengeful and powerful people, but this place was inside another type of arena, an arena formed by a crowd who were camped outside the building of the major office of WPU. News about vandals breaking into TV channel studios and media group's offices made these powerful people rethink their next decisions — especially since most of the armed forces refused to follow any orders against Antony and the crowd who was supporting him. The only way WPU's members had to calm the population down was to choose a popular leader who everyone could trust.

The President of the United of States, Richard Sowell, became very popular after his role in the Nebraska battle was revealed to the world. Nobody imagined that a president in a wheelchair would take part in a battle and also lead a huge squad of aircraft. Also, his actions were interpreted by the population as heroism, and his leadership was seen as the main reason many soldiers were saved — which helped Antony save humanity from certain extinction. On the other hand, Yelena, who spent months accusing Antony of causing everything bad on the planet, was put away and kept hidden from the media by WPU's members — because she became the most hated person worldwide.

Antony still didn't know what he felt when he saw President Sowell. Relief, admiration, doubt. It was obvious that he'd missed a lot of information and news about what was happening in the world while he was in jail, but despite that, he was sure that Richard Sowell wouldn't act one hundred percent according to his principles. There are many requirements for politicians, governmental leaders and businessmen to satisfy, it didn't matter if Antony and Richard had saved humanity. Also, most of these people around the defendant, despite the new leader of WPU, had unbreakable egos. However, they were intelligent enough to proceed, dodging the limits ahead.

Having the people break into WPU's major office would be fatal, but with Richard Sowell there as a leader, it would be less probable.

In this scenery full of rumors and conspiracies, the President of the United States and new leader of WPU started his statement in a calm but confident voice.

"Dear ladies and gentlemen of this honorable intergovernmental organization, it's hard to describe in a few words the honor and pride I feel to be in front of this select group. I'm not here in the name of the country of which I was elected president nor any other country. I'm here in the name of all people who live on this planet. I know that sounds prepotent, but I swear... I do not feel this way. Today I and the honorable members of this intergovernmental organization are here to affirm our decision on the case of the destruction of Cloud Base's building and its net technology, which was property of the Sims family, by its own heir and one of the main owners, Antony Keith Sims. Certainly, one of the great inventions of humanity in the last centuries, which made it possible for the population to no longer suffer facing the death..."

Antony looked like the statue of a saint as he watched the President speaking. Behind this articulated speech there were reasons, opinions, wills, and decisions from everyone here. So, if Antony was a defendant of President Sowell, he was a hostage in this arena. Also, it was a surprise for Antony to remember he had a middle name.

"... According to the Twentieth Treaty of Vienna signed by the members of this intergovernmental organization thirty-one years ago, all residents of Cloud system were considered humans beings. Therefore, the defendant present in this assembly has been charged with killing exactly four billion eighty-five million ninety-two thousand five hundred and three people. Based on these numbers and comparing them with those from others murderers, most of them terrible dictators, Antony Keith Sims is considered by ninety-nine percent of WPU as the perpetrator of the greatest genocide in human history, and there is only one fate that this honorable intergovernmental organization defined..."

The greatest genocide in human history... well, it's a pompous name to fill school books, Antony thought, knowing that this fate chosen for him would be a worthy condemnation for his *crime.*

"However, the members of this organization don't have full power to decide the life of the defendant. There is still the opinion of the people outside and mine."

Whispers sounded louder in the chamber. That was the evidence that the last words from Richard Sowell belonged to him and not to the *hostage* of WPU's members.

"I still remember your actions that day in New York and our last talk, a talk which no one here knew about. Also, I really hope one day you have the opportunity to meet my granddaughter, Lucy, who was born a month ago. Being a grandfather is a dream that I had waited for a long time. I don't have scientific evidence that you are the responsible for this, but I just wanted to get it off my chest."

The whispers sounded louder, and the doubtful faces multiplied around the arena.

"Anyway... everyone here knows their limits, including me and the huge crowd outside," Richard's voice sounded out loudly. "Summing it up, we are here to do our best, and our best is respecting the lives who passed away and the lives who stay close to us. The decision of WPU evaluated and signed by me is... the defendant Antony Keith Sims is sentenced to life imprisonment in a supermax prison which will be chosen in a few days by a special commission of diplomats."

A wind of hope hit Antony, but he contained his feelings, unlike the people around him who just expressed surprise and indignation.

Richard Sowell took a deep breath to calm his voice down and finished his speech, giving the only chance for the defendant to speak about the verdict.

"Antony Sims... it's your time. Do you have any question or plea about the verdict of WPU?"

Antony, showing calm and doubt, approached the microphone in front of him.

"Humm... Can I at least have a beer fridge in my cell?"

Voices of indignation and screams like "oh, no" echoed through the luxurious assembly as Richard Sowell answered, smiling.

"Of course, my friend."

Antony just replied with a smile.

EPILOGUE

Six years later — Copenhagen, Denmark.

A next-generation, matte black Tesla slowed down on a deserted street as it approached the main gate of the Vestre prison — a cross-shaped, ancient building converted into a high security prison seven years before thanks to a special inmate.

It was a beautiful sunny day of summer in Copenhagen; the blue sky was clear, and the green of the lawns and trees was vivid, as the tranquil landscape was dominated by silence.

The Tesla was parked close to the main gate, and from it, a tall woman dressed in an elegant, dark gray suit walked out. She stood for a while, watching the fences around the huge cross. The woman had an athletic body and wide shoulders covered by her long black dreadlocks.

The prison fences were covered with photos, flower arrangements, plush toys, and candles — All types of homage to the main inmate of the prison, including posters and letters with phrases like *thanks for saving my family*, *my kids are here because of you,* and *thank you, because of you I could become a mother* — each more loving than the last.

Suddenly, something flew from behind the fence up to the sky. It was a white dove, which drew the woman's attention. *It looked like an angel,* she thought, as if in a daydream. Then she continued on her way.

She crossed the large gate and entered the prison. Inside, in a modern and wide office where some prison officers were working, she signed several documents. Then she was searched. Also, she asked the officers something like, *Is everything ready in the backyard?* After several protocols, she was taken to the next office with armored, heavy double doors.

Through a wide and long corridor, the woman crossed seven offices with heavy doors, and in each of them, she was submitted to the same ritual: signing several documents and being searched.

Finally, she crossed the last office and arrived at a well-lit, huge room that resembled a library. In the back of the room, there were several metal bookshelves filled with all kinds of book titles. Close to the last shelf, there was a simple and old spinet piano covered with some sheets of music. In the middle of the room, there was a large rustic table covered with many manuscripts and stacked books, surrounded by some simple chairs. At the table, a thin guy with a shaved head dressed in a dark blue inmate uniform was focused, writing on papers. He was engrossed that he didn't notice the presence of the visitor who stood for a while, just looking around. Additionally, a detail in this room impressed the woman: two big beer fridges next to the small library were filled with many kinds of drinks. After noticing that, the woman finally spoke:

"Hey bro, your situation is not so bad here."

Antony calmly raised his head and saw Glenda, looking completely different from what he remembered.

"Wow, Glenda... you look different. To be honest, you look like an angel."

"You look different too, Antony... to be honest, hmmm... Ok, you don't look like an angel."

"Hahahaha..."

Both laughed, then Antony got up and went to hug his friend who had tears rolling down her face. They lost sense of time in a long hug that had taken seven years to happen.

"My friend... I thought you had forgotten about me."

"Little chicken... it's impossible for anyone to forget you in this world."

They acted like their last meeting had been a week before, despite Glenda having faced many bureaucracies for six years before she was authorized to see her friend.

Antony didn't look so surprised, but he was excited about his first visit since he had been arrested in this special jail, where he had been reading all kinds of books, playing the piano, and drinking alcohol. With the exception of his haircut, Antony's appearance hadn't changed. After such a long time without talking to anyone, he felt a little pain in his jaw but not so hard to stop his constant smile.

After this long hug, they stepped back and looked at each other again.

Glenda looked gorgeous, dressed in designer clothes and wearing light makeup that highlighted her beautiful black skin. She seemed stronger and taller than the last time she had met Antony. Also, her dreadlocks made her look empowered, but her captivating smile broke the ice of her serious appearance.

Both took a seat close to the table.

"Hey, look at you, Glenda. You're not fixing robots anymore, I guess."

"Sometimes, just for fun. Now, I'm Ms. Jackson, an attorney ready to defend the downtrodden people. Well... I know I won't save the world like you, but it's the best I can do."

"An Attorney! Wow... It's amazing, seeing you so well. You really deserve all the best from this world... besides, I didn't save it," Antony said, smiling the whole time, "I just did what I thought was the right thing to do. But I don't wanna talk about it. Hey... grab a drink from the fridge. I have a lifetime supply of beer... Also, what about the folk in the community... and your son, William?"

"Hey, hey, let's break it down, bro... first, I don't drink when I'm working, thanks... furthermore, it's amazing to see you again. Well... the community went back to the same place. Now it's more beautiful, like a fairy tale village and with kids too. They have made our community shine," Glenda's look was heartwarming, "William is ok, studying to be a mechanic engineer. He's not so good with numbers, but he's a hardworking boy."

"Nice... very nice, and a relief too. Since I got here, I haven't had access to any information from outside. I was afraid something bad might have happened to your son since... you know, that battle," Antony calmed down.

"The world outside... the world that you saved, is ok too. Far from being perfect, but better than you remember. It's hard to take some people away from their powerful positions, but me and many other good people are always persevering."

"It's good to know that... sometimes I think the world outside is over and just me and my books remained here. To be honest, everything you see here — food, drinks, books — I have received through an entrance that looks like a pet door. While I receive this stuff, I know the world outside is still alive."

"It sounds so lonely."

"I got used to it... despite missing some people, I realized that living alone has always been in my nature. My mind never stops here. By the way... no more Angels flying out there?"

"After you made Cloud vanish... no more Angels. Just some videos recorded on phones and security cameras inside Cloud Base, which people on the internet insist on having arguments about whether they are real or fake. The fact is that these creatures became legends."

Glenda noticed a disappointment on Antony's face as he looked away.

"Do you miss her?"

"Are you asking about the woman you took your hairstyle from?"

"Your sense of observation changed nothing... and yes, I am."

"Our last talk, in a Cloud Base's infirmary, wasn't very... friendly. I just wanted to say sorry and thanks for guiding me. She was the true hero. And your meeting with her was probably great."

"She saved my life and said I deserve all blessings of the universe... I'll never forget her."

For a moment, Glenda got a little emotional and looked away as she remembered her only and unforgettable meeting with Muriel. After a few seconds, she came around.

"So... like I just said... everything outside is going well despite attempts to build another Cloud. There are more schools, security, hospitals, churches, and temples of several religions. To me, the most important thing is that people have improved what is really necessary... they have improved their mentality independently from the government, ideologies with bad intentions, or media. The end of Cloud and the noise you made was important for people to wake up to reality."

"It sounds magical, but another Cloud? It's insane. It was probably Yelena's idea. She's tireless," Antony looked indignant.

"Yes, you're right. She and her puppets. Now she has been still, hidden in some low position in the German government. Something like an *assistant of an assistant*. But as you just said, she's tireless and she can make trouble anytime," Glenda sounded a little angry. "You won't believe, after our first battle in Nebraska, she kept my son, William, in an induced coma just... just to blackmail me... so I would lead the soldiers around the damn Cloud Base. My son was alive the whole time, and she insisted that he was on the verge of death. Oh gosh, it drove me up the walls."

"Hey, hey... calm down," Antony approached her, "the worst is over. According to what you said, your son and the world are doing well.

This kind of people, like Yelena, must be watched forever... unfortunately."

"I see, sorry. But I still remember our last talk. I still remember her telling me about how she could manipulate genetics and control the natality of humanity... controlling genetically the races... who would be stronger or more intelligent... Oh gosh, I can feel the fear just like when she told me that all over again."

Antony took a deep breath.

"It's unbelievable how people like that can exist, but they can. Also, she hid something very important from you. All babies who were created in laboratories didn't survive for more than two years. Every one of them was born with a mysterious kind of anomaly, a failure in their neural connections. Something that had no cure. I saw some of them when I worked with my family." Antony looked very serious, "It was insane and monstrous. This was one of the reasons I left my position at Sims Corp. I saw some of these babies and... their brains didn't connect to their organs. They looked empty, like —"

"Like they didn't have souls."

"Yes... like they didn't have souls. The end of humanity was inevitable if what happened hadn't happened."

Glenda noticed some sketches among several manuscripts on the table. She took one of them in her hand and saw drawings of angels and giant robots, which made her remember the big adventure years ago. Beyond the manuscripts, a sign under the sketches, T.S., drew her curiosity too.

"T and S? Hmmm... does it mean something special? You look busy here."

"It's my signature, it means Tony Sims. Sounds like the name of a hero, hehehe... don't you agree? And all these papers... well, it's our story, all about me, my family, you, the community, about Cloud, the greedy interests of Sims Corp... and Angels. There are many secrets and details to show to the world. Information that can't vanish. Making sketches helps me clear up my ideas."

Glenda took other papers as she kept talking to Antony.

"It looks like a book of memories of someone who is about die. You're still so young," Glenda smiled, "and have time enough to write lots of books here."

"Maybe, if I die, I won't go to a virtual heaven that my grandfather built anymore. You remember the mess the members of WPU, my family, and other bad people made in the world," Antony looked calm and serious. "So, the people must know the truth about what they did so that no one else tries to destroy the place they live again. Everything is already written here on this table. It will be a brave new book. Now, I just need to review and organize the pages."

"Antony, thanks to you, humanity can have kids again. Your mission is over; you don't need to worry anymore. I struggled a lot to become an attorney, and my goal is to get you out of here. It's unfair." Glenda looked worried, "you sacrificed your good life to change the world. Now you are isolated here and —"

"Why do you think they gave me books, good meals, beers, and a piano? They want to keep me isolated from the world but not demented. These bad people you have faced still want me to create a new Cloud or something, but I won't. Like you just said, we must defend the downtrodden people... wherever we are."

"But there are people outside who need you and —"

"NO... people outside need the truth. It doesn't matter if you're a prisoner or an attorney, if you're on a battlefield or sweeping the sidewalk in front of your home... What I mean is, we live in an eternal war between good and evil... This war can be noisy, with guns and violence, or it can be silent, through damned bad ideologies that destroy our sense of what's real, but the fact is... we will always be in this war. And what can we do? Be good and do good... Be good, respecting others, in your neighborhood, in schools, in our workplace, being attentive to old people, kids, animals, everyone. Be good doing your job honestly and doing your best. Be good fighting against the bad people, acting with justice and ethics... So, the people outside don't need someone to tell them what is the truth; they just need the truth, and they get it from being good... every single day."

"Glenda looked down, trying to find words to say, as Antony showed the calm of a priest."

"I thought... I thought you wanted to get out of here."

"Yes, I do. I want it so much. And we both know well that's very hard. But... that's the world where we live."

It was hard for Glenda to admit that Antony was right, and that a perfect world with her friend set free would be a faraway reality.

Suddenly, she noticed on one of the sketches in her hands the title, written in large letters, created by Antony for his book. The title was mysterious and she tried to change the subject.

"*Against the World*, that's the title of your new book?"

"Yes, it's an amazing title. Don't you like it?"

"Hmmm... I don't know, yet... It sounds so dramatic."

"Do you have any other idea? Maybe *Tony in Wonderland*. I can change it because of you."

"Hahahaha... No, there's no need. This is a pretty cool title. Hey, I have a surprise for you. Get your fat ass out of that chair and follow me out of here."

"Out of here?" Antony looked suspicious, "how did you get that?"

"Just follow me, bro. We don't have much time. Come on."

Antony followed Glenda, and they crossed the gate that was still open. Then they walked into a long, brightly lit corridor made out of concrete and steel, with no doors. At the end of the corridor, a light shone intensely, and Antony saw the silhouette of a short person who looked like a woman.

"Who is there? And where are we going?"

"I told you, it's a surprise."

"It's the second person I get to meet in six years. I hope it's a good surprise... hey... it can't be!"

He started to walk faster, away from Glenda, as tears rolled down his face. Antony still wasn't sure who the person at the end of the corridor was, but little by little, his suspicions were confirmed. He ran towards them, his heart beating faster than it had ever beaten in these last few years, and an almost forgotten feeling, called happiness, took over this moment.

"My mom... I missed you... I missed you so much."

Antony hugged tight the short woman who was waiting for him with arms wide open.

"My son... I missed you so much, too... Oh, my bright boy."

It seemed like time hadn't passed for Nivedita. She was the same loving woman whom Antony considered his mother in reality while Angela lived in Cloud. Antony lost all sense of time again in a long hug that he had waited for since he escaped from his own penthouse apartment in Brussels.

"Let me look at you, bright boy. I like your new haircut. And you don't look so thin. Have you been eating well here?"

"Yes, I have, Nive... well, I live in a prison. There isn't much to do here except for eating and sleeping."

Tears ran down their faces, reaching their smiles.

"Hey, how are you? And your family? Is everything ok with them? Do you need help with anything?"

"Hahaha... the same guy, worried about me. Calm down, my bright boy," Nivedita talked slowly, like a careful mother. "Me and my family are fine. Ms. Jackson has helped me a lot these last years. She became my close friend, like your mother had been in the past. Come on, let me show you something."

Nivedita took her son by his arm, then they walked slowly to the backyard of the prison.

"Hey, what's... what's this?"

"More surprises, of course."

The sunlight was vivid, and to Antony, who had lived six years away from it, everything looked blurry for a moment, until he saw many familiar people on an improvised soccer field. He needed some seconds for his brain to reason and to remember each one of them. Everyone was wearing sports clothes and looking very happy.

A very outgoing, strong, blond man was the first to approach him, and Antony almost couldn't believe who he was.

"Aaron? It's insane... What are you doing here?" Antony looked impressed.

"Hi, Mr. Sims... Antony is better, I guess. Hahaha... I'm glad you remember me."

"Of course, I still remember when I threatened to turn off your airplane mid-flight."

"Yes, I remember that hahaha... and you were inside it too. It scared me to death."

"Wow... I hope it didn't cause you any traumas."

"Trauma? That was the best thing that ever happened to me. After that, I left my job as Yelena's assistant. I always hated that damn cow. So, I went back to my former job as a soccer coach, and I'm still working with that... thanks to you. Hahaha... come on."

Aaron hugged Antony tight, as he looked embarrassed, surrounded by the strong arms of his new smiling friend. Then he noticed another strong man approaching and recognized him immediately.

"Wow... Who is this guy? He is strong like a buffalo... William."

"Hey, how are you? You look like the same little chicken just like my mother said hehehe... I'm joking, bro... sorry."

"No worries, bro... After six years of silence, any joke sounds like a symphony. Come on and give a hug, too."

The two friends hugged each other. Then Antony saw the rest of the people in the back. Everyone who had been in the community when he visited and played soccer with seven years ago was there. Girls and boys had become beautiful, strong, and happy adults. Antony hugged each one warmly — and with each hug he remembered good and funny moments from the few days he had stayed in the frontier of Kansas and Nebraska. Smiles, jokes, and good memories took over this moment.

Glenda and Nivedita stood up, just looking at everyone, enjoying a good moment like proud mothers watching their grown-up kids.

When the people calmed down, Antony asked about the soccer field they were standing in.

"Hey, what's the meaning of this field?"

"Come on... don't you remember?" William replied.

Suddenly, Antony could remember the *fateful* soccer game where he was forced to play in the community — the game through which he'd got Glenda's help to protect Cloud Base in Nebraska. The dimensions of field were very similar to the community's field with two soccer goals, that were 3 ft x 2.5 ft, on each side.

"Let me refresh your memory," Glenda said smiling, "you made a bet seven years ago with me and these guys. Remember? If you won the game, then I would help you. Unfortunately, you passed out before the end of the game. So, they're all here to finish it."

"Hey, but I scored the goal? Didn't I?"

"I don't know. If you hadn't passed out, you would remember... hahaha... Actually, I will never tell you if you scored the goal or not."

"Wow... that's unfair. But let's play this insane game hahaha..."

The guys separated into two teams of six players, one led by William and the other by Antony. Aaron was the referee, and Nivedita, Glenda, and the rest of the guys stayed around the field, supporting the teams of their choice.

Antony called the guys of his team to gather in a circle and talked about their strategy for the game.

"Ok, guys, this is my game plan to win. And the plan is *play without a plan.* Just have fun. So... HERE WE GO!"

Both teams positioned themselves on the field as Aaron set the soccer ball between them. After he blew the whistle, euphoria took over the field. The guys played and ran frantically while the supporters, including Glenda and Nivedita, clapped and talked loudly.

"Come on, William. Prove to this little chicken's team who is the boss on the field."

"Hey... the team of my bright boy can kick William's team's ass anytime."

"Hey... take it easy, Nivedita, hahaha..."

Both ladies had a lot of fun watching the game.

Two guys from William's team knocked the ball around until it entered the area and was tackled by Antony and one of his teammates. Then Antony gained ball control and played a well-timed pass to his forward mate, who took his shot and hit the post.

"Hey bro, you're not so bad for a prisoner," William said out loud.

"Despite my knees aching, it's alright. It happens after six years inside a jail, hahaha... Let's go, bros."

Shoots, passes, crosses, scores, screams, laughs, happiness... all ingredients for a fun soccer game among friends.

Antony was very excited, running frenetically all over the field, smiling and calling shots to everyone, all without realizing that his heartbeats were increasingly faster.

"Come on, bright boy... all that's missing is you scoring a goal."

"Yes, little chicken. Move your fat ass faster and score the goal that you owe me."

"Come on, girls, just give me some more time hahaha..."

The shouts from Glenda and Nivedita sounded like an order to Antony, who still hadn't felt his heart beating faster than normal. His consciousness seemed to be disconnected from his body while he played like a child having fun on a playground — like there was no tomorrow.

A foul made the game stop. So, Aaron restarted the game in favor of Antony's team.

The defender lost no time and passed the ball quickly to the midfielder who played a perfectly-placed pass to Antony. He was between the center circle and the box. Also, nobody was marking him.

"GOOOO, ANTONY, IT'S YOUR CHANCE!"

Antony had possession of the ball and sprinted toward the soccer goal. His eyes gleamed brightly, his smile seemed to contain all the happiness in the world, and his heart continued beating fast… so fast that it suddenly stopped completely. Nevertheless, he kept running, smiling, and full of enthusiasm.

He kicked the ball toward the soccer goal and kept running without paying attention to whether the ball got in the goal. But he kept running and smiling. He raised his hands and head up to the sky as the excited people behind him clapped.

Antony didn't realize that his heartbeats had stopped; he kept running. He looked up, feeling the intensity of the sunshine and a deep inner peace. Suddenly, he saw a dove flying slowly over him, in front of the sun. *It's an angel… an angel*, he thought.

His heart stopped completely, but he kept running, smiling, with his hands up as the people behind him looked excited. Then, everything became white in front his eyes, up in the sky.

It was the first time in his life that Antony felt free.

Cezar Marcilio is a Brazilian author with degrees in History and Art, constantly creating new projects and seeking ways to share his ideas with the world. His major influences include the works of Michael Crichton, Dan Brown, George Orwell, Isaac Asimov, and the renowned philosopher Olavo de Carvalho. In his debut book, Haven't We Met Before?, he explores the consciousness of individuals in their pursuit of goals and passions. In Against the World, Cezar pushes the boundaries of his imagination, blending sci-fi and fantasy to tell a story about how a select group of people can manipulate society to their advantage, potentially causing harm to our planet and the universe.

Currently, Cezar lives with his family in his homeland, where he works as an Art teacher.

Check out some of the other titles from Robb Publishing:

Made in United States
Troutdale, OR
11/03/2026

56844522R00246